MATERIALS SCIENCE AND TECHNOLOGIES

# LOW-DENSITY POLYETHYLENE

## PROPERTIES AND APPLICATIONS

# MATERIALS SCIENCE AND TECHNOLOGIES

Additional books and e-books in this series can be found on Nova's website under the Series tab.

MATERIALS SCIENCE AND TECHNOLOGIES

# LOW-DENSITY POLYETHYLENE

## PROPERTIES AND APPLICATIONS

JOHAN GEISLER
EDITOR

Copyright © 2020 by Nova Science Publishers, Inc.

**All rights reserved.** No part of this book may be reproduced, stored in a retrieval system or transmitted in any form or by any means: electronic, electrostatic, magnetic, tape, mechanical photocopying, recording or otherwise without the written permission of the Publisher.

We have partnered with Copyright Clearance Center to make it easy for you to obtain permissions to reuse content from this publication. Simply navigate to this publication's page on Nova's website and locate the "Get Permission" button below the title description. This button is linked directly to the title's permission page on copyright.com. Alternatively, you can visit copyright.com and search by title, ISBN, or ISSN.

For further questions about using the service on copyright.com, please contact:
Copyright Clearance Center
Phone: +1-(978) 750-8400      Fax: +1-(978) 750-4470      E-mail: info@copyright.com.

## NOTICE TO THE READER

The Publisher has taken reasonable care in the preparation of this book, but makes no expressed or implied warranty of any kind and assumes no responsibility for any errors or omissions. No liability is assumed for incidental or consequential damages in connection with or arising out of information contained in this book. The Publisher shall not be liable for any special, consequential, or exemplary damages resulting, in whole or in part, from the readers' use of, or reliance upon, this material. Any parts of this book based on government reports are so indicated and copyright is claimed for those parts to the extent applicable to compilations of such works.

Independent verification should be sought for any data, advice or recommendations contained in this book. In addition, no responsibility is assumed by the Publisher for any injury and/or damage to persons or property arising from any methods, products, instructions, ideas or otherwise contained in this publication.

This publication is designed to provide accurate and authoritative information with regard to the subject matter covered herein. It is sold with the clear understanding that the Publisher is not engaged in rendering legal or any other professional services. If legal or any other expert assistance is required, the services of a competent person should be sought. FROM A DECLARATION OF PARTICIPANTS JOINTLY ADOPTED BY A COMMITTEE OF THE AMERICAN BAR ASSOCIATION AND A COMMITTEE OF PUBLISHERS.

Additional color graphics may be available in the e-book version of this book.

## Library of Congress Cataloging-in-Publication Data

Names: Geisler, Johan, editor. Title: Low-density polyethylene : properties and applications / Johan Geisler, editor.
Description: Hauppauge : Nova Science Publishers, [2020] | Series: Materials science and technologies | Includes bibliographical references and index. | Summary: "Low-Density Polyethylene: Properties and Applications examines the rheology of low-density Poly(ethylene)-based systems. Processing this commodity, alone or in combination with different micro/nano-fillers, requires a deep knowledge of its rheological behavior in order to set up the process parameters. Following this, the comprehensive research progress on low-density polyethylene is reviewed, and the mechanisms of low-density polyethylene biodegradation are summarized. Additionally, the effect of microorganisms on low-density polyethylene and products of this degradation with their level of toxicity is discussed. Later, the authors focus on the different types of low-density polyethylene, microorganism-mediated degradation, changes in the physiological properties of low-density polyethylene post degradation and its applications in other fields. The detailed knowledge of preferential sorption is studied in an effort to reveal new information regarding low-density polyethylene properties. Consequently, the usage of low-density polyethylene in membrane separations is promoted"-- Provided by publisher.
Identifiers: LCCN 2020028235 (print) | LCCN 2020028236 (ebook) | ISBN 9781536181920 (paperback) | ISBN 9781536182736 (Adobe PDF)
Subjects: LCSH: Polyethylene. Classification: LCC TP1180.P65 L69 2020  (print) | LCC TP1180.P65  (ebook) | DDC 668.4/234--dc23
LC record available at https://lccn.loc.gov/2020028235
LC ebook record available at https://lccn.loc.gov/2020028236

*Published by Nova Science Publishers, Inc. † New York*

# CONTENTS

| | | |
|---|---|---|
| **Preface** | | vii |
| **Chapter 1** | Rheological Behavior of Low-Density Poly(Ethylene) and of Its Composites<br>*Rossella Arrigo and Giulio Malucelli* | 1 |
| **Chapter 2** | A Comprehensive Study on Biodegradation Process of Low Density Polyethylene by Microorganisms<br>*Seyyed Mojtaba Mousavi, Maryam Zarei, Seyyed Alireza Hashemi, Wei-Hung Chiang, Ahmad Gholami and Yasin Sadeghipoor* | 55 |
| **Chapter 3** | Biotransformation Rendered by Microbes on Various LDPE<br>*Anushree Suresh and Jayanthi Abraham* | 81 |
| **Chapter 4** | Membrane Separations Using Low-Density Polyethylene Membranes<br>*Alena Randová, Lidmila Bartovská and Karel Friess* | 115 |
| **Index** | | 139 |

# PREFACE

*Low-Density Polyethylene: Properties and Applications* examines the rheology of low-density Pply(ethylene)-based systems. Processing this commodity, alone or in combination with different micro/nano-fillers, requires a deep knowledge of its rheological behavior in order to set up the process parameters.

Following this, the comprehensive research progress on low-density polyethylene is reviewed, and the mechanisms of low-density polyethylene biodegradation are summarized. Additionally, the effect of microorganisms on low-density polyethylene and products of this degradation with their level of toxicity is discussed.

Later, the authors focus on the different types of low-density polyethylene, microorganism-mediated degradation, changes in the physiological properties of low-density polyethylene post degradation and its applications in other fields.

The detailed knowledge of preferential sorption is studied in an effort to reveal new information regarding low-density polyethylene properties. Consequently, the usage of low-density polyethylene in membrane separations is promoted.

Chapter 1 - This chapter reviews the current state of the art referring to the rheology of low-density Poly(ethylene) (LDPE)-based systems: undoubtedly, processing this commodity, alone or in combination of different micro- to nano-fillers, requires a deep knowledge of its rheological behavior, in order to be able to set up the process parameters. In particular, after an overview of the fundamentals of rheology of thermoplastics (also including the main rheological tests employed for this characterization), the chapter will discuss the specific rheological behavior of LDPE, and how fillers ranging from micro- to nano-scale may affect it.

Chapter 2 - In today's life polyethylene can be introduced as a resistant polymer to degradation. Since it has several biological and chemical properties it plays a critical role in the preparing process of different products like plastic bags. Polyethylene types can be classified into several groups such as Cross-Linked Polyethylene (XLPE), Linear Low-Density Polyethylene (LLDPE) High-Density Polyethylene (HDPE) and Low-Density Polyethylene (LDPE). Their density, number of branches and surface functional groups are different which can provide numerous applications. LDPE can be achieved by a high-pressure approach which creates short and long chains. Also, it can be stated that LDPE possesses a translucent structure with 50 to 65% crystallinity. From an economic point of view, LDPE is a polymer with acceptable flexibility, enough strength and satisfactory cost which can widely be applied in drug and food packaging. Regarding the environmental issues associated with LDPE waste, many efforts have accomplished in order to identify different microorganisms for degradation process. The present chapter will cover different topics, first the comprehensive research progress of LDPE is reviewed, and second the mechanisms of LDPE biodegradation are summarized. After that, different aspects like bio and oxo-degradable LDPE are presented. In two last parts, the effect of microorganisms on LDPE and products of this degradation with their level of toxicity are discussed respectively. Finally, to boost the improvement for LDPE, the major opportunities and future challenges are also explained.

Chapter 3 - The Indian plastic market is facing enormous challenges from industries, technology and environmental organisations. The surging increase in the waste generated calls for newer biotechnological waste management techniques. There is an exponential increase in the usage of plastic commodities particularly low-density polyethylene. In the coming years the consumption of LDPE is expected to escalate further. There have been various conventional methods followed for recycling of plastic solid waste. The reuse of plastics will alleviate the problem caused by plastics in the environment one such successful example is usage of recycled LDPE for greenhouse application having excellent mechanical and physical properties and resistant to weathering. This book chapter focuses on the different types of low-density polyethylene, microorganism mediated degradation, different changes in the physiological properties of LDPE post degradation and its application in other fields. Microbial mediated degradation of LDPE has led the scientist to investigate the different metabolic pathways involved in the degradation process. The microbial mediated degradation of LDPE is initiated by the secretion of extracellular or intracellular enzymes thereby resulting in the cleavage of long polymer chain compounds to monomers. Also, a detailed explanation on the effect of fungal activity on LDPE is provided in this chapter.

Chapter 4 - Low-density polyethylene (LDPE) belongs to the family of thermoplastic materials. LDPE is formed from the ethylene monomer units and its density is generally low due to the branching from the main chain and due to the lower portion of solid, impermeable crystalline parts. Properties of LDPE, such as stability, resistance and toughness, make this material suitable for various applications, especially in the packaging industry (containers, bottles, laboratory equipment, bags etc.). Nevertheless LDPE sorbs some amounts of gases and liquids. Therefore, the study of LDPE sorption phenomena of various penetrants constitutes an important issue. The detailed knowledge of the preferential sorption, i.e., the sorption of one component of the mixture, together with the total sorption of material under specific conditions and penetrant compositions can reveal new information regarding LDPE properties. And, consequently, the usage of LDPE in membrane separations is promoted.

In: Low-Density Polyethylene
Editor: Johan Geisler

ISBN: 978-1-53618-192-0
© 2020 Nova Science Publishers, Inc.

*Chapter 1*

# RHEOLOGICAL BEHAVIOR OF LOW-DENSITY POLY(ETHYLENE) AND OF ITS COMPOSITES

*Rossella Arrigo and Giulio Malucelli*
Politecnico di Torino, Dept. Applied Science and Technology and
Local INSTM Unit, Alessandria, Italy

## ABSTRACT

This chapter reviews the current state of the art referring to the rheology of low-density Poly(ethylene) (LDPE)-based systems: undoubtedly, processing this commodity, alone or in combination of different micro- to nano-fillers, requires a deep knowledge of its rheological behavior, in order to be able to set up the process parameters. In particular, after an overview of the fundamentals of rheology of thermoplastics (also including the main rheological tests employed for this characterization), the chapter will discuss the specific rheological behavior of LDPE, and how fillers ranging from micro- to nano-scale may affect it.

## 1. INTRODUCTION

Rheology is a branch of physics that describes the mechanical behavior of materials during flow-induced deformation [1-2]. Rheological studies are not specifically focused on ideal elastic materials or ideal fluids, the behavior of which can be described by the well-known Hooke and Newton models, respectively. Conversely, rheology often focuses on materials that are able to exhibit elastic, viscous or both behaviors under different flow conditions [3-6]. This kind of materials, referred as *complex fluids*, include polymers, gels, emulsions, foods, biofluids or inks. For this reason, in the last decades the rheology has become an important field in materials engineering, food science and biotechnology [7].

As far as polymeric materials are concerned, in the melt state they exhibit a rather complicated and unusual flow behavior; unlike water, oil or organic solvents, polymers are non-Newtonian fluids [8-10]. By definition, fluids deform when a force is applied and continue to deform until the force is removed; in a Newtonian fluid, the rate of deformation is directly proportional to the applied force [11]. Conversely, polymers in the molten state do not exhibit a direct relationship between the rate of deformation and the stress applied to the melt, hence showing a non-Newtonian response [12]. More specifically, polymeric materials exhibit a so-called viscoelastic behavior, involving both a viscous and an elastic component; additionally, the response of a polymer upon deformation is time dependent [13-14]. All these issues make rheology a very useful tool for characterizing polymer systems.

Besides, rheological characterization has a primary role in polymer research, being a fundamental link between the production of polymers and their end-use properties. In fact, especially for thermoplastic polymers, the knowledge of the flow behavior is essential for all production processes, as they typically involve the melting of the material, its subsequent shaping through the flow in a die or the filling of a mold and, lastly, its solidification into the final product [15-17]. Since the fundamental part of the productive process occurs while the polymer is in the molten state, the processing of thermoplastics is determined by their flow behavior, which in turn depends on both polymer structure and selected processing conditions, in terms of applied temperatures, pressures and stresses [18-20]. In this view, the rheological characterization is firstly necessary for the design of the processing equipment; as an example, the evaluation of the dependence of viscosity as a function of the applied shear rate at different temperatures allows assessing the flow behavior of polymer melts [21]. In Figure 1, a typical viscosity curve of a linear polymer is shown, along with the different shear rate ranges corresponding to the common processing operations for thermoplastics [22-23]. It is evident that the rheological behavior of thermoplastics and their processing are strictly related; the knowledge of the polymer flow behavior is, thus, mandatory to model and design their processing. Besides, a deep understanding of the polymer rheological characteristics enables to solve process troubleshooting and to optimize the processing conditions [24].

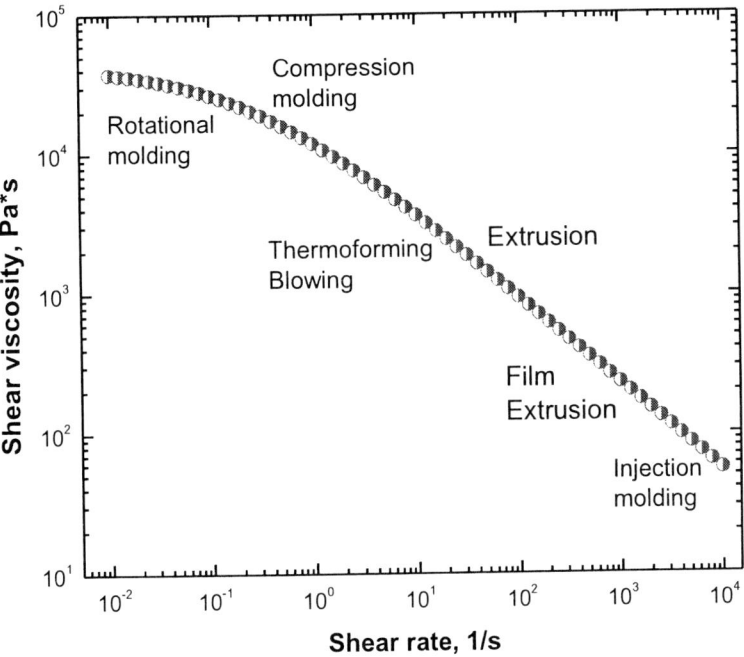

Figure 1. Processing shear rates.

On the other hand, the rheological properties are very sensitive to the microstructural characteristics of polymers, such as molecular weight, molecular weight distribution and presence of either short or long chain branching [25-27]. More specifically, the flow behavior is strongly affected by the structural architecture of polymer chains, since the presence of high molecular weight chains or long branches is able to modify the motion dynamics of macromolecules, leading to changes in the polymer relaxation spectrum [28]. In this context, the study of the polymer rheological behavior can be profitably exploited to gain some insight into the chemical structure of the material and to characterize its branching structure that is difficulty assessed through classical spectroscopic or chromatographic techniques [29].

Additionally, as far as polymer-based composites and nanocomposites are concerned, the rheological characterization is a powerful tool to infer the state of distribution of micro- and/or nano-sized fillers, as well as the possible occurrence of strong polymer-filler and filler-filler interactions [30-31]. In fact, the rheological response of polymer-based complex systems reflects their intimate microstructure: more specifically, when the formation of interconnected structures of nanofillers, such as percolative networks, occurs, the motion of long polymer chain segments is restricted, thus resulting in a change in the relaxation spectrum of the material [32-34]. Rheological measurements are hence capable of revealing fundamental information as far as the microstructural evolution of filled polymers is concerned, being very sensitive to changes in composite internal structure and to the relaxation dynamics of polymer chains [35].

This chapter aims to thoroughly describe the rheological behavior of low-density Poly(ethylene) and the correlation occurring between its rheological properties, molecular structure and melt processing. Before detailing this part, the basic flow characteristics of thermoplastics and the typical rheological measurements performed on polymers will be discussed. Afterward, the focus will be devoted to the main rheological characteristics of LDPE, also considering the possible influence of micro- and nano-fillers on the polymer viscoelastic response.

## 2. FUNDAMENTALS OF POLYMER RHEOLOGICAL MEASUREMENTS

### 2.1. Shear Flow

For the determination of the polymer viscosity in a shear flow field, the so-called two-plate model, sketched in Figure 2, is used. In this model, the polymer is trapped in between two parallel surfaces: the bottom plate is fixed, while the upper one moves at constant speed.

If the no-slip boundary conditions are satisfied, the velocity of the fluid varies linearly from $0$ at the bottom to $V$ at the upper surface; the linear velocity profile generates a constant velocity gradient, known as shear rate ($\dot{\gamma}$) [36]. The shear viscosity ($\eta$) of the polymer is defined as the resistance that the fluid offers to the shear deformation, and can be derived from the ratio between the imposed shear stress $\tau$ (i.e., the force $F$ tangentially applied, divided by the area $A$ of the plate) and the shear rate $\dot{\gamma}$ [37]:

$$\eta = \frac{\tau}{\dot{\gamma}} \tag{1}$$

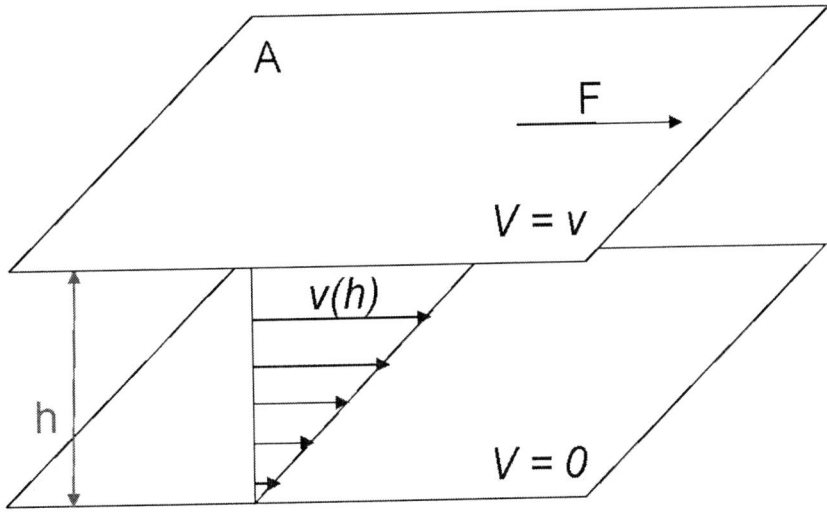

Figure 2. Schematic of simple shear flow between two parallel plates.

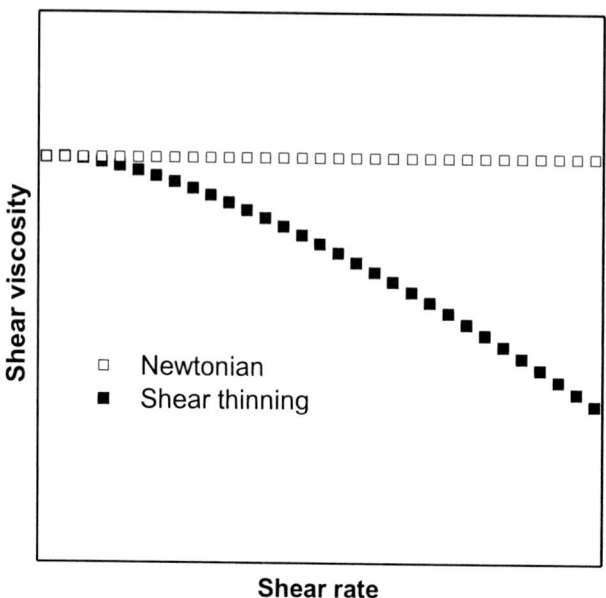

Figure 3. Flow curves for Newtonian and shear-thinning fluids.

This relation is known as Newton's Law of Viscosity; if the viscosity is independent from the shear rate, the fluid exhibits Newtonian behavior and its flow curve (i.e., the plot of the shear viscosity as a function of shear rate) is a straight line [38]. Conversely, molten polymers show a non-Newtonian behavior, involving a dependence of the viscosity from the applied shear rate. In particular, polymers are "shear-thinning" or pseudo-plastic fluids, since their viscosity decreases as the shear rate increases [39]. In Figure 3, the typical flow curves for Newtonian and shear-thinning materials are reported.

For thermoplastics, the shear viscosity approaches the ideal Newtonian behavior at low shear rate values: the region, in which the viscosity is constant, is called "Newtonian plateau" and the values of η in this zone is the *zero-shear* or *Newtonian* viscosity ($\eta_0$) [40]. As the shear rate increases, the viscosity starts to decrease, passing through a transition region towards the shear-thinning zone, in which a dramatic drop of the viscosity values can be observed. This decrease is due to the occurrence of two concurrent phenomena: the preferential alignments of the macromolecules along the flow direction and the disentanglements of the long polymer chains upon the applied deformation [41]. The higher the shear rate, the easier is the deformation of the polymer; this finding implies that the flow of the polymer through dies or other process equipment is facilitated at high shear rate values.

Different models are commonly employed for describing the flow behavior of pseudo-plastic fluids. The simplest viscosity model is the Power-law model, requiring two fitting parameters:

$$\eta = K\dot{\gamma}^{n-1} \qquad (2)$$

where $K$ is the consistency of the fluid, accounting for the magnitude of the viscosity, and $n$ is the power-law coefficient [42]. This model is able to describe only Newtonian plateau (n = 1) or shear-thinning region (n < 1), but is of less use for polymers showing a pseudo-plastic behavior.

An extension of the power law is the Cross model:

$$\eta = \frac{\eta_0}{1+(\lambda\dot{\gamma})^{1-n}} \qquad (3)$$

which introduces the following fitting parameters: $\eta_0$ = zero-shear viscosity; $\lambda$ = characteristic relaxation time; $n$ = power-law coefficient [43].

For certain thermoplastics, a better fit is achieved using the Carreau model, showing the same fitting parameters as the Cross model [44]:

$$\eta = \eta_0[1+(\lambda\dot{\gamma})^2]^{\frac{n-1}{2}} \qquad (4)$$

Rheological measurements in shear flow for thermoplastics are usually carried out using rotational or capillary rheometers (Figure 4). The first typology includes plate-plate and cone-plate configurations and the deformation of the material is performed through the mutual rotation of the two plates [45]. Rotational rheometers are able to work in strain-controlled or stress-controlled mode, depending whether the imposed variable is the strain rate or the stress, respectively. The polymer, in the form of granules or as a disk-shaped sample (preliminary obtained through compression or injection molding), is placed in between the plates and, after reaching the thermal equilibrium, is sheared while the torque or the deformation are collected as a function of the angular velocity of rotation or stress, respectively. Typically, these devices allow an accurate temperature control along with a high torque resolution, but they present a limitation about the maximum achievable deformation rates or stresses. Differently, the use of capillary rheometers enables to reach shear rate values similar to those experienced by the polymer during an injection molding or an extrusion process; for this reason, the capillary rheometers (Figure 4C) are usually used to measure the shear viscosity at high shear rate ranges [46]. To this aim, polymer granules or powder are firstly fed in a pre-heated barrel and then extruded through a capillary die at a specific piston speed; the viscosity function of the material is derived from the measurement of the melt pressure at the entrance and within the die. However, due to the complex flow field in the capillary, the obtainment of accurate viscosity values requires several corrections [47].

### 2.1.1. Dynamic Oscillatory Shear Tests

Dynamic oscillatory measurements represent the most common method to measure the flow behavior of a thermoplastic material using a rotational rheometer [48]. In this test, the material is subjected to a sinusoidal stress or strain and the resulting mechanical response is collected as a function of time. More specifically, the sample is oscillated about its initial equilibrium position in a continuous cycle: the amplitude ($\gamma$) of the applied oscillation corresponds to the maximum stress or strain and the angular frequency ($\omega$) represents the number of oscillations per second. As regards an ideal elastic material, for which the stress is proportional to the strain, both strain and stress are in-phase; conversely, a purely viscous fluid shows 90° phase difference between stress and strain, since in this case the stress is proportional to the derivative of the strain. Polymer melts exhibit a viscoelastic behavior, i.e., partly elastic and partly viscous: therefore, their mechanical response falls between the two extremes, showing a phase difference between stress and strain ranging from 0 to 90° [49].

In the case of thermoplastics, the ratio between the applied stress (or strain) and the measured strain (or stress) gives the complex modulus G*, a quantitative measure of the polymer resistance to the deformation [50]. The complex modulus can be decomposed into the in-phase and out-of-phase components, representing the elastic and the viscous characteristics of the polymer, respectively.

More specifically, G* can be calculated as:

$$G^* = G'(\omega) + iG''(\omega) \tag{5}$$

where G' (storage modulus) is the in-phase and G" (loss modulus) is the out-of-phase component of the polymer viscoelastic behavior. G' represents the elastic part, being related to the energy stored in the material, while G" stands for the viscous feature of the polymer response and refers to the energy dissipated in the deformation [51].

The ratio between the two moduli is the loss factor

$$\tan\delta = G''/G' \tag{6}$$

where δ is the phase angle between stress and strain.

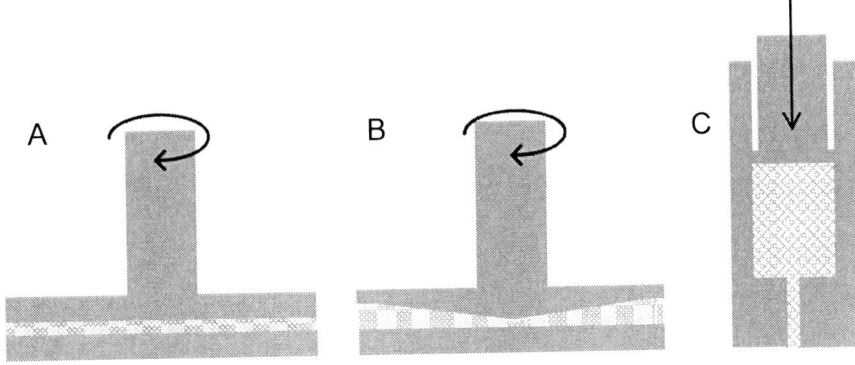

Figure 4. Schematics of plate-plate (A), cone-plate (B) and capillary (C) rheometers.

Furthermore, the complex viscosity of the polymer can be derived through a dynamical measurement as follows:

$$\eta^* = \sqrt{(\eta')^2 + (\eta'')^2} = \sqrt{(G'/\omega)^2 + (G''/\omega)^2} = G^*/\omega \qquad (7)$$

Dynamic oscillatory shear tests can be performed either in the linear or nonlinear viscoelastic region of the polymer. Figure 5 schematically depicts the results of an isothermal strain sweep test, in which the response of the material is recorded as a function of the strain amplitude and a fixed frequency [52]. At low strain amplitude values, the linear viscoelastic regime can be detected: it is characterized by the independence of both moduli from strain amplitude. In the linear region, the applied stress or deformation are low enough to cause structural breakdown, and the oscillatory response remains sinusoidal. As the applied strain amplitude is increased, a transition from linear to nonlinear regime occurs, indicated by the dramatic drop of both moduli. In the nonlinear region, the moduli are strain amplitude-dependent; besides, the resulting periodic response is distorted and diverges from a sinusoidal wave [53].

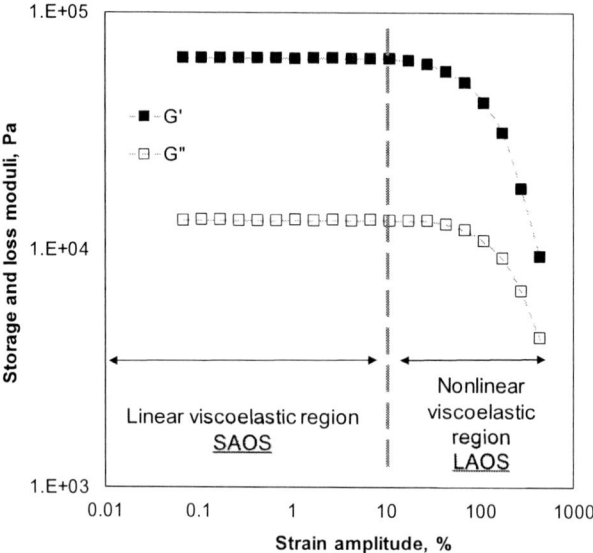

Figure 5. Storage and loss moduli measured in a typical strain sweep measurement.

Small amplitude oscillatory shear (SAOS) measurements are carried out within the linear viscoelastic region of the material, providing useful information about the microstructure of polymers and polymer-based complex systems [54]. In Figure 6, the typical results of a SAOS test for a thermoplastic are reported; the measurement is performed at constant temperature and strain amplitude (low enough to be in the linear viscoelastic region of the material), by varying the frequency in the range $10^{-2}$ - $10^{2}$ rad/s. From the analysis of the complex viscosity curve, the Newtonian plateau at low frequencies can be observed, followed by the shear thinning region, characterized by a rapid decrease of the viscosity values as a function of the oscillation frequency. As far as the moduli curves are considered, a monotonic increasing trend as a function of frequency can be observed; in the low frequency range (the so-called terminal region), the polymer melt is predominantly viscous (G" > G') and both storage and loss moduli curves exhibit a frequency dependence typical of a liquid-like rheological behavior (G' α $\omega^2$ and G" α ω). In this region, the rheological response of the material is governed by the relaxation processes of long chain macromolecules. As the frequency increases, the polymer behavior becomes mainly elastic (G' > G") and the viscoelastic response is governed by the fast dynamics of short polymer chains [55].

Although a SAOS test is able to accurately describe the flow behavior of a complex fluid, being based on a rigorous theoretical foundation, the linear viscoelastic region extends for quite small strain amplitude values. However, in a typical process operation, the polymer is subjected to large and rapid deformation; SAOS characterization is not hence sufficient to fully evaluate the flow behavior of the polymer in practical applications and it is necessary the study of the viscoelastic response of complex fluids in the nonlinear region through large amplitude oscillatory shear (LAOS) tests [56]. The measurement typically used to describe the type of nonlinear behavior is the strain sweep, depicted in Figure 7. Generally, four kinds of strain amplitude dependence can be observed: type I or strain softening, in which both moduli decrease as a function of strain amplitude; type II or strain hardening, in which both moduli increase; type III or weak strain overshoot, in which G" shows an overshoot; type IV, or strong strain overshoot, in which both moduli show overshoot. Type I behavior is typical of unfilled polymer melts, where the macromolecules tend to align along the flow direction for high strain amplitude; otherwise, the overshoots present in type III and IV are usually related to the formation of weakly interacting structures in gels or composites [57]. Additionally, the strain dependence of the rheological functions in the nonlinear region is helpful in characterizing the branching degree of polymer chains.

## 2.1.2. Stress Relaxation Tests

A typical stress relaxation test involves the application of an instantaneous deformation to the polymer sample, and the successive monitoring of the stress decay as a function of time [58]. This kind of measurement, that is called also step strain, allows evaluating the shear modulus, defined as:

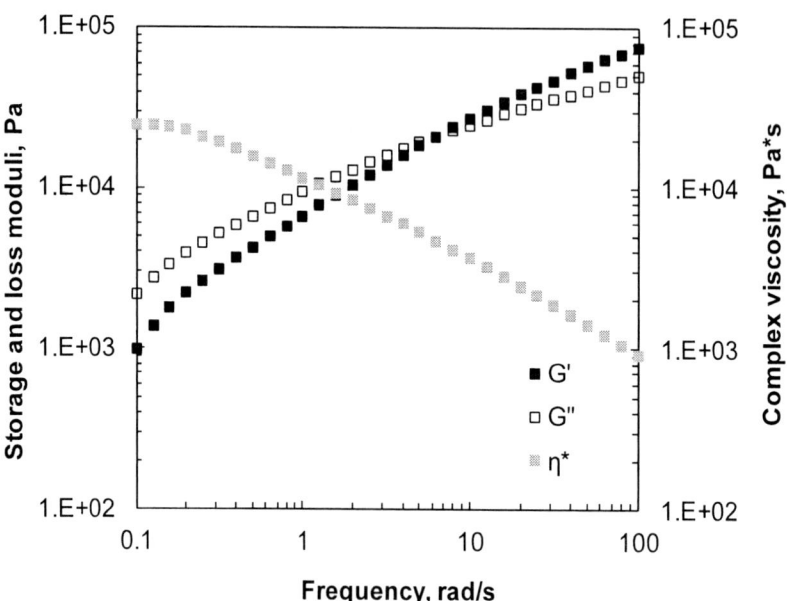

Figure 6. Typical results of a SAOS measurement.

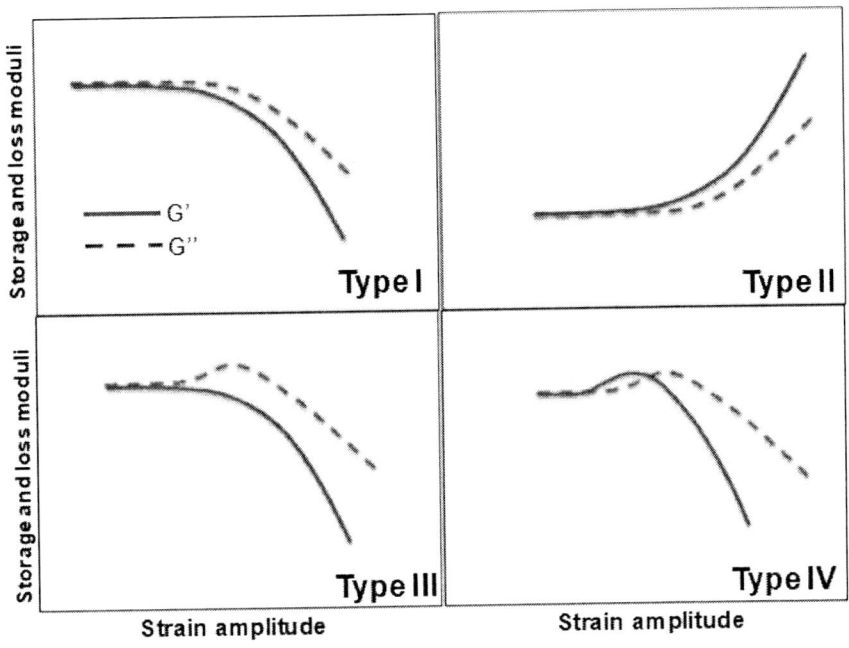

Figure 7. Different kinds of nonlinear rheological behavior.

$$G(t, \gamma_0) = \frac{\sigma(t, \gamma_0)}{\gamma_0} \quad (8)$$

Therefore, shear modulus is a function of time and strain amplitude; however, if the strain amplitude value is low enough to be in the polymer linear viscoelastic region, the dependence of G(t) from $\gamma_0$ is negligible, and the modulus depends only on time. In this case, the shear modulus is called relaxation modulus and represents the decay of the stress as a function of time, after the deformation of the polymer. For unfilled polymers, G(t) tends to reach a value of equilibrium equal to zero after a certain time; materials showing this behavior are viscoelastic liquids, which dissipate all the energy provided with the deformation. Differently, materials with nonzero equilibrium modulus are defined as viscoelastic materials, since they are able to partly store the energy applied on the sample [59].

Stress relaxation modulus can be exploited to classify polymeric materials depending on their relaxation dynamics. Polymer melts usually shows the aforementioned terminal behavior, typical of liquid-like systems. For these materials, the stress relaxes completely, and G(t) approaches zero after a certain time once the stress is applied. Conversely, crosslinked materials or polymer-based nanocomposites, in which the formation of a percolative network of nanofillers occurs, exhibit a solid-like rheological behavior. For these systems, the presence of bridges between the macromolecules or the establishment strong polymer/filler and filler/filler interactions, make more difficult the motion of the polymer chains, slowing down their dynamics and avoiding their complete relaxation [60].

## 2.2. Elongational Flow

In many relevant processing operations, polymer melt is subjected to the elongational flow, i.e., a stretching deformation that, depending on the specific process, can be uniaxial or biaxial [61-63]. As an example, elongational flow plays a key role in fiber spinning, blow molding, film blowing or foaming. Additionally, the elongational deformation of the melt also occurs in some processes dominated by shear, such as extrusion and injection molding; the change of the die diameters or the injection of the polymer into the mold gate, represent some examples, in which the melt experiences stretching deformations [64].

In analogy to the shear flow, the elongational viscosity ($\mu$) is defined as the ratio between the applied stress and the deformation rate [65]:

$$\mu = \frac{\sigma}{\dot{\varepsilon}} \qquad (9)$$

In Figure 9, the typical trend of the uniaxial elongational viscosity as a function of time is reported for different values of deformation rate. In steady-state conditions, if the system is in its linear range of deformation, the so-called Trouton law applies, and the time-dependent elongational viscosity is three times the time-dependent shear viscosity [66]:

$$\mu_0(t) = 3\eta_0(t) \qquad (10)$$

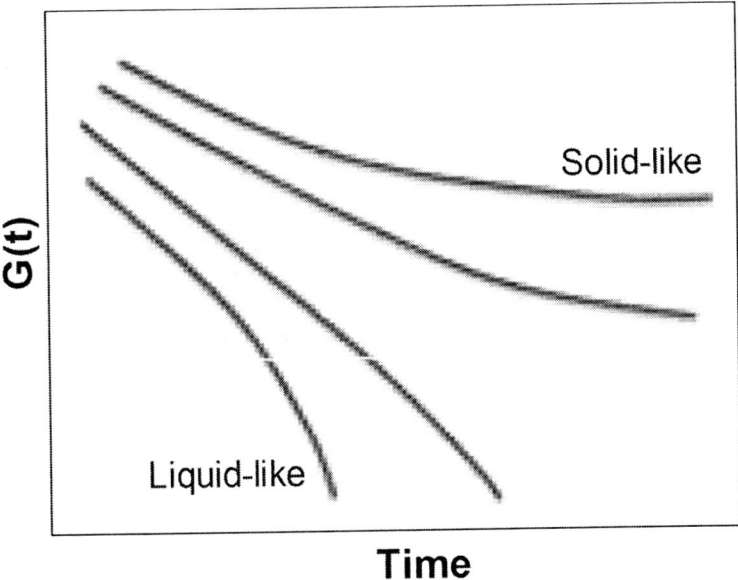

Figure 8. Stress relaxation behavior of different classes of polymers.

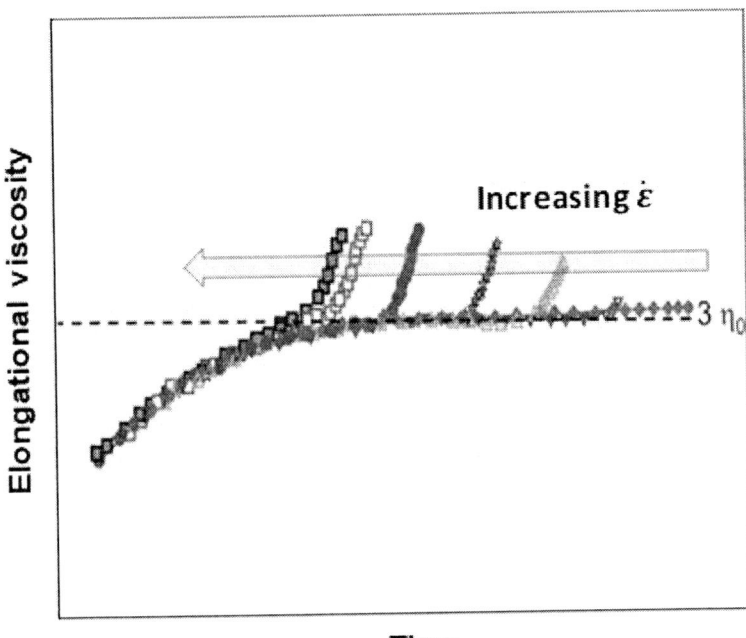

Figure 9. Elongational viscosity at different rates of deformation.

Looking at the curves shown in Figure 9, the elongational viscosity increases as a function of time, reaching a plateau corresponding to the Trouton value for low deformation rates. Differently, at higher rates, the elongational viscosity exhibits a steep upturn that occurs at progressively shorter times as the deformation rate increases. This peculiar behavior, due to the growth of the stress as a result of the strain increase, is called strain hardening and is beneficial for the homogeneous deformation of the polymer in processing operations like film blowing or foaming [67-68].

Elongational viscosity can be used to describe the extensional behavior of polymers in isothermal conditions; however, during the most common processing operations, in which elongational flow is involved, the temperature of the material changes while processing. Therefore, to fully evaluate the flow behavior of polymers in practical real applications, rheological characterization in non-isothermal elongational flow is performed [69]. The commonly used equipment is provided with a series of pulleys, which grab the hot polymer filament coming out from an extruder, and deliver it into a final pulley rotating at steady acceleration, while a load cell measures the force on the filament [70]. At the breaking of the filament, the current force and the speed of the final pulley are recorded. The fundamental properties derived from such an experiment are the melt strength (MS) and the breaking stretching ratio (BSR). MS refers to the force in the molten polymer at breaking, while BSR represents the maximum elongation of the melt and is calculated as the ratio between the drawing speed at breaking and the extrusion velocity [71].

## 3. RHEOLOGICAL BEHAVIOR OF LDPE

As stated in the Introduction, the rheological properties of a polymeric material, both in shear and in elongational flow, strictly depend on the structure of the polymer, in terms of molecular weight, distribution of molecular weight and presence of short and long branches. In the following, the main feature of the flow behavior of low density poly(ethylene) will be discussed, considering the relationships between the rheological functions and the macromolecular structure of this polymer.

## 3.1. Shear Flow Behavior

### 3.1.1. Effect of the Molecular Weight

The shear viscosity as a function of the shear rate of a series of LDPE with different molecular weight ($M_w$) has been evaluated by Acierno et al. [72]; the molar mass distribution of the investigated samples and their degree of branching are very similar, while their weight average molecular weight significantly differ, ranging from $67*10^3$ to $166*10^4$. Two different effects can be observed as a result of the increase of the polymer $M_w$: a significant growth of the zero-shear viscosity ($\eta_0$) values in the low shear rete region, and a progressively more pronounced shear thinning behavior. In other words, the zero-shear viscosity becomes remarkably higher as $M_w$ increases, but the differences in the viscosity values tend to diminish at high shear rate.

Generally, as far as the enhancement of the zero-shear viscosity is concerned, the well-known Mark-Houwink equation can be used to predict the effect of the polymer $M_w$ [73-74]:

$$\eta_0 = K_1 M_w \quad for \quad M_w < M_c \tag{11}$$

$$\eta_0 = K_2 M_w^\alpha \quad for \quad M_w > M_c \tag{12}$$

where $K_1$ and $K_2$ are parameters depending on the kind of polymer and temperature, α lies between 3.4 and 3.6 and $M_c$ is the critical molar mass. In Figure 10, this relation is applied to determine the critical molecular weight of a series of LDPE with different $M_w$ [75].

Several reports discuss about the possible dependence of the Mark-Houwink equation on the molar mass distribution [76-78]. In this context, Stadler et al. [79] evaluated through linear viscoelastic measurements, the zero-shear viscosity of twenty-four LDPE samples with different $M_w$, characterized by polydispersity index ranging from 1.8 to 16. The obtained data were represented in a $\eta_0$ vs. $M_w$ plot, in order to verify the applicability of the Mark-Houwink equation for all materials. The results showed that all the experimental data follow the trend predicted by the power-law relation, indicating its independence from molar mass distribution. This finding allowed using the plot of zero-shear viscosity as a function on the molar mass to discriminate polymers with different macromolecular architectures; in fact, as the Mark-Houwink equation is independent from the polydispersity index of the polymer, any deviation from this relationship can be attributed to the existence of complex molecular architectures, such as tree-like or star-like structures [80].

Wood-Adams et al. [81] performed SAOS measurements of seven samples of LDPE with the aim to investigate the effect of the molecular weight on the polymer linear viscoelastic behavior. The $M_w$ of the selected samples ranges from $38*10^3$ to $33*10^4$, while their polydispersity index is approximately 2. From the analysis of the flow curves, an increase of the zero-shear viscosity values and an amplification of the shear thinning as a function of $M_w$ occur. Due to the high molecular weight of the selected materials, the calculation of $\eta_0$ from the viscosity curves was not possible, and the viscosity values were determined using discrete relaxation spectrum. The fitting of the experimental data with the Mark-Houwink equation confirmed the validity of the used procedure, as the fitting parameters coincide with those already reported in literature for LDPE [82]. Additionally, the effect of the molecular weight on the loss angle was studied, documenting a significant influence of the polymer molar mass on the frequency, at which the elastic feature of the LDPE viscoelastic behavior becomes prominent. Furthermore, important information about the calculation of the molecular weight between entanglements were obtained from the analysis of the storage modulus G' as a function of $M_w$ with increasing the polymer molar mass, a progressive disappearance of the terminal behavior was observed, due to the higher number of entanglements in the high molecular weight LDPEs.

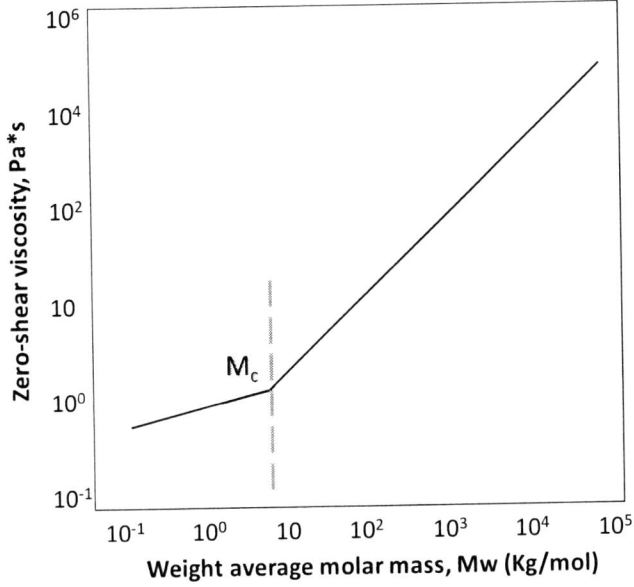

Figure 10. Flow curves of different LDPE samples with different $M_w$ (A) and zero-shear viscosity as a function of $M_w$ (B).

As far as the influence of the molar mass distribution on the flow behavior of LDPE is concerned, it has been shown that the breadth of the distribution has a quite irrelevant effect on the flow curve shape. In fact, LDPE samples characterized by different polydispersity index, exhibited a very similar trend of the complex viscosity as a function of frequency, and the differences in the molar mass distribution caused a slight broadening of the transition zone of the curve, at intermediate frequencies between the Newtonian plateau and the shear thinning region [83-85].

An interesting study by La Mantia et al. [86] was addressed to derivate a relationship between the LDPE molecular weight and the so-called die-swell, i.e., the property related to the swelling ratio of a polymer filament coming out from a capillary. From a technological point of view, the die-swell is of fundamental importance to predict the diameter of a polymer filament exiting from an extrusion die, being the ratio between the actual diameter of the extrudate and the die diameter. The existence of this swelling phenomenon in thermoplastics is due to the elastic component of their viscoelastic behavior, since the polymer is able to expand at the exit of a capillary because of the stored normal stresses. Usually, the experimental measure of the die-swell is difficult, as the extrudate may freeze at the exit of the die, before reaching its maximum recovery; additionally, in the case of low viscosity polymers, the falling of the exiting strand under its own weight needs to be considered. Therefore, a prediction of this property from the polymer rheological behavior should help in solving many practical issues. Six different LDPE samples, differing for $M_w$ and polydispersity index were selected and characterized through shear measurements. The obtained data showed an increase of the swelling ratio with increasing both molecular weight and molecular weight distribution. Besides, an analytical expression for the calculation of die-swell was found, taking into account the effect of the polymer molecular structure, the geometrical characteristics of the used dies and the operating conditions, as well.

Besides die-swell, a further issue commonly encountered during the extrusion of high molecular weight LDPE is the occurrence of the so-called sharkskin phenomenon, i.e., the appearance of instabilities during the processing resulting in periodic irregularities on the extrudate surface [87]. The sharkskin phenomenon is usually observed for extrusions at high shear rate (high production rates) and is attributed to the cohesive rupture of the polymer at the die exit, due to the rearrangement of the velocity profile at the die tip. More specifically, if the imposed tensile stress exceeds the cohesive strength of the polymer, a rupture of the extrudate surface occurs; for this reason, this phenomenon is also called melt-fracture [88]. Several reports documented the effect of the molecular characteristic of the polymer on the appearance of this phenomenon, showing an increase of the critical shear stress as a function of the breadth of the molecular weight distribution [89-91]. Furthermore, Yamaguchi et al. [92] documented a decrease of the critical stress for the onset of sharkskin as a function of the increased molecular weight between entanglements. Interestingly, Ansari et al. [93] found a relationship between the onset of the sharkskin and the plateau modulus, implying that LDPEs with higher modulus experience this phenomenon at higher shear rates and stresses.

The described relationships between linear viscoelastic functions and LDPE molecular weight were often used as an alternative to the classical chromatographic methods, to obtain some information concerning the polymer $M_w$ and molecular weight distribution [94-95]. The use of rheology instead of gel permeation chromatography has some advantages: first, as many commercial LDPE samples with high $M_w$ do not dissolve in solvents at room temperature, only the devices operating at high temperature are able to perform the analysis. Furthermore, rheological functions such as viscosity or storage modulus can be measured in a more convenient way than GPC elution curve. For all these reasons, several procedures have been proposed, such as viscosity methods, involving the use of Mark-Houwink equation [96], or empirical correlations using storage modulus [97], among a few to mention.

### *3.1.2. Effect of the Macromolecular Architecture*

The topology of the macromolecular chains, in terms of presence of short or long branches anchored to the main polymer backbone, has a relevant effect on the rheological functions of thermoplastics in general, and particularly for LDPE [98]. In general, there are three different kinds of branch architectures: star structure, with a certain number of branches radiating from a center; comb structure, with uniformly spaced side-chains and random branching structure, which is the most probable architecture for LDPE obtainbered through radical polymerization [99]. For LDPE containing random distributed branches, a further classification between short and long chains branching needs to be taken in account, since it is of fundamental importance as far as their influence on the viscoelastic properties of polymer is concerned. In fact, from a rheological point of view, long chain branching (LCB) are the side-chains long enough to improve the entanglement number of the polymer [100]. In this case, the introduction of LCB is hence able to introduce novel relaxation modes related to the motion of these long chains, which in turn can bear further minor ramifications, leading to an amplification of the non-Newtonian feature of the polymer flow behavior [101]. On the other hand, the presence of LCB can also reduce the hydrodynamic volume of the polymer [102]; therefore, these two concurrent phenomena having an opposite effect on the polymer rheological behavior need to be considered to fully understand the effect of the macromolecular architecture in affecting polymer viscoelastic properties. Conversely, the presence of short chain branching (SCB) has less impact on the rheological functions, as the density of entanglements remains almost unchanged with respect the correspondent linear polymer [103].

Numerous investigations have been performed concerning the effect of SCB and LCB on the rheological properties of several LDPE samples having different chain topology. Several investigations [104-106] report a lower melt viscosity for LDPE as compared to a linear polyethylene with similar $M_w$, although usually an impressive increase of the viscosity values was observed for LDPEs having low density of LCB and high molecular weight. Bersted at al. [107] documented a different effect of LCB depending on the branching mechanism; in particular, a more pronounced effect on the polymer rheological behavior was found for LCB introduced through peroxide-initiated reactions with respect to those by thermal initiation. Different samples of LDPE characterized by different densities of LCB, similar $M_w$ and narrow molecular weight distribution, were synthetized by Yan et al. [108] using a high temperature/high pressure process and a titanium-based catalyst. Compared to a linear polymer having the same molecular characteristics, LDPE samples exhibited progressively higher zero-shear viscosity and lower viscosity at higher shear rates as a function of LCB content. Furthermore, the increase of the LCB content caused an amplification of the elastic feature of the polymer viscoelastic behavior, with an increase of the relaxation time and a higher level of die-swell.

Wood-Adams et al. [81] demonstrated a remarkable effect of the LCB density on storage modulus and loss angle. More specifically, the storage moduli curves as a function of frequency showed the presence of a shoulder in the low frequency region, associated with the appearance of a new well-defined relaxation regime in the branched polymer that is not present in the linear counterpart. Furthermore, they found that loss angle is a very sensitive indicator of the LCB presence. As far as the loss angle trend is concerned, linear sample exhibits the typical terminal behavior, approaching loss angle = 90° at low frequencies; conversely, in the curve of LDPE sample, a plateau at intermediate frequencies appears, the extension and magnitude of which depend on the LCB density.

Stadler et al. [109] analyzed the viscosity functions of several LCB polyethylenes, fitting the obtained data with a Carreau-like model involving 6 fitting parameters to determinate the characteristic relaxation time of selected polymers. They pointed out that a proper evaluation of the dependence of the rheological functions on LCB level is possible only if both $M_w$ and LCB concentration were exactly known, since in many cases the influence of long-chain branches and the molar mass could not be separated.

The rheological behavior of a LDPE suitable for film-blowing was investigated and compared with that of linear and SCB polyethylenes [110]. According to other reports, the presence of LCB enhanced the non-Newtonian features in the trend of the viscosity curve (i.e., disappearance of the Newtonian plateau and amplification of the shear thinning behavior). Besides, the LDPE sample exhibited higher values of die-swell as compared to the others materials, because of its more prominent elastic behavior.

Valenza et al. [111] presented a set of rheological data for a series of LDPE polymerized in presence of different comonomers, differing in $M_w$ and in the length of grafted branches. The general trend of the collected data suggested that the non-Newtonian behavior in shear flow is significant dependent on the length and on the density of LCB. Furthermore, the critical shear rate, at which surface irregularities (sharkskin) appear during extrusion was evaluated, showing that the activation energy for this phenomenon is strongly affected by both $M_w$ and LCB structures.

The presence of LCB also affects the thermorheological behavior of thermoplastics, i.e., the temperature dependence of the relaxation mechanisms occurring in polymers. In this context, Stadler et al. [112] analyzed the thermorheological behavior of various LDPE samples, proving their thermorheological simplicity; this finding implies that the flow activation energy for these samples is independent from the relaxation time. Furthermore, activation energies were measured from the relaxation spectra of selected materials, in good agreement with the literature.

## 3.2. Elongational Flow Behavior

The rheological behavior of LDPE in elongational flow has been widely investigated, being this polymer largely processed through different processing operations, such as film-blowing or thermoforming, in which uniaxial or biaxial deformation is prominent. Results from extensional experiments performed on a standard LDPE at constant elongational rate $\dot{\varepsilon}_0 = 0.1\ s^{-1}$ [113] demonstrated the typical elongational behavior of LDPE, characterized by a steep increase of the stress at the beginning of the measurement, followed by a region, in which the stress-strain curve shows a growing slope (strain hardening behavior) and then a plateau for $\varepsilon_H > 4$.

Wolff et al. [114] studied the elongational properties of a commercial LDPE sample through creep and creep-recovery experiments, aiming at assessing the differences in the flow behavior between linear and nonlinear range of deformations. First, they found that the breadth of the linear range of LDPE is higher in elongational than in shear flow; besides, the experimental results indicated that, in the linear range, the Trouton law is fulfilled, implying that the value of the elongational viscosity is three times higher the value of the zero-shear viscosity. In the nonlinear range of deformations, as expected for long branched polymers, LDPE exhibited strain hardening behavior, with a rapid growth of the elongational viscosity as a function of stress. Münstedt et al. [115] formulated several blends based on LDPE in order to improve the film blowing properties of a series of linear polyethylenes. Virgin LDPE shown the usual strain hardening behavior as a function of time and the increase of viscosity was found more pronounced for the higher elongational rate probed. Conversely, for linear polyethylene samples the strain hardening was not observed, and the viscosity values, after a nearly constant regime, decreased. The addition of 10 wt.% of LDPE to the linear polymer resulted in the appearance of a defined strain hardening at high elongational rates; though at low rates this behavior was not detected, the elongational viscosity values do not drop down, indicating that the sample remains homogeneous during deformation. Furthermore, the homogeneity of the produced blown films was assessed through the measurement of the film thickness at 32 positions along the bubble take-up direction, and the calculation of an "inhomogeneity index". The virgin LDPE film showed a very low value of the index, which remained constant during the entire duration of the characterization, reflecting the general favorable effect of strain hardening on the uniformity of deformation; conversely, for the linear polymer, higher values of the index were obtained, with significant data scattering during the measurement. The blend containing LDPE exhibited an improved stability of the film thickness, although it did not reach the performance of virgin LDPE.

Aiming at assessing the relationship between elongational viscosity and molecular structure, Münstedt et al. [116] characterized three LDPE samples differing for $M_w$, molecular weight distribution and LCB content. In the linear range of deformation, all the investigated samples exhibited a very similar behavior, and the values of their elongational viscosity coincided with the Trouton value. Differently, in the nonlinear range the viscosity curve is shifted towards higher values as a result of the increase of Mw and of a broadening of the molar mass distribution. In a similar way, also the presence of LCB affected the shape of the elongational viscosity curve, causing a progressive increase of the viscosity maximum in the nonlinear range of deformation, as a function of the degree of branching.

Two samples of LDPE were synthesized in a laboratory-scale autoclave under high pressure, and their elongational behavior was compared to that of two commercial materials [117]. Preliminary GPC analyses performed on the synthesized LDPEs revealed the presence of high molecular weight tails, resulting in a bimodal molar mass distribution. Rheological characterization in elongational flow showed that these samples exhibit improved strain hardening behavior as compared to the commercial samples, because of the longer characteristic relaxation time of the fraction of high molecular weight macromolecules.

As far as the rheological behavior in non-isothermal elongational flow is conceded, La Mantia et al. [118] evaluated the MS and BSR of different LDPE samples having different $M_w$ and polydispersity index. Additionally, to evaluate the influence of the chain topology, one sample of LDPE was obtained in a vessel reactor, showing "three-like" branching, while all the other samples were synthesized in a tubular reactor. An increase of the polymer $M_w$ causes an increase of the melt strength and a decrease of the polymer deformability for all LDPE samples, except for the polymer with three-like branching, which exhibited higher MS and lower BSR as compared to the expected values. Therefore, the presence of a different chain topology caused a decrease of the polymer stretchability and an increase of the melt resistance. This behavior was explained considering that the presence of three-like structures makes more difficult the orientation of the macromolecules along the flow direction, causing a drastic reduction of the polymer extensibility and an increase of the melt resistance.

## 4. EFFECT OF FILLERS ON THE RHEOLOGICAL BEHAVIOR OF THERMOPLASTICS: GENERAL REMARKS AND SOME APPLICATIONS TO LDPE

The scientific literature reports several papers dealing with the rheological behavior of filled thermoplastics, but only few are specifically focused on LDPE-based systems.

### 4.1. Nanoclays

Clay nanofillers (i.e., nanoclays) are nanoparticles of layered mineral silicates. It is well-known [119] that their inclusion in a polymer matrix can lead to different morphologies, which derive from the aptitude of macromolecular chains to enter the clay galleries (the so-called *intercalation* phenomena) or even to collapse the stacked structure of clays by delaminating their sheets (*exfoliation* phenomena). Besides, when no interactions take place between the organic and inorganic constituents, phase separation occurs and the resulting composite structure is considered as *separated phase*: as a consequence, the so-derived polymer/clay composites are classified in the range of traditional micro-composites.

Generally speaking, the rheological behavior of clay composite systems that do not exhibit intercalation or exfoliation phenomena is similar to that of the polymer matrix, due to the low filler loadings and the weak interactions taking place between the two phases [120, 121]. Conversely, the nanocomposites exhibiting intercalated or exfoliated morphologies are primarily prone to develop a particular rheological behavior [122] that comprises the following phenomena:

- a very low nanoclay loading (usually below 5 wt.%) is enough for promoting a transition between quasi-liquid and quasi-solid behavior. This transition determines a significant increase of G' in the low frequency region
- the protracted application of large deformations produces a significant decrease in the linear viscoelastic module and the disappearance of the quasi-solid state of the nanocomposite system.

The transformation from a quasi-liquid to an almost solid behavior has been mainly interpreted on the basis of the formation of a percolation network, which occurs at very low nanoclays loadings because of their high anisotropy [123, 124].

Some of the advances on the rheology of LDPE-clay systems will be summarized in the following. The rheological effects of LDPE containing Cloisite Na⁺ (i.e., a typical montmorillonite clay), first modified with octadecyl ammonium chloride and then with equimolar amounts of octadecylamine (with respect to the pristine modified clay) were thoroughly investigated and correlated with the level of intercalation achieved as a consequence of the nanoclay modifications. Besides, a commercially-available organo-modified clay (namely Cloisite 30B) was treated with the same modifiers and compounded with LDPE in a Brabender mixer unit at 3 wt.% loading [125]. The complex viscosity of the systems based on LDPE and its nanocomposites was not affected by the presence of the modified nanoclays, because of the occurrence of limited intercalation phenomena. Conversely, when the modified nanoclays were added to LDPE-EVA (14 w/w of vinyl acetate) compounds, higher basal spacing values were achieved, hence increasing the complex viscosity of the nanocomposite system with respect to the unfilled counterpart.

Two different clays (namely, montmorillonite and kaolinite) were compounded in LDPE, using a twin-screw extruder and two screw configurations with different shear intensities. In particular, one configuration was set for intensive mixing and high residence time, while the other for more conventional, lower residence time compounding processes [126]. The state of nanoclay dispersion was also investigated through rheological analyses carried out with a rotational rheometer: in particular, at low frequencies, none of the tested samples, comprising unfilled LDPE, showed a Newtonian plateau. Besides, the complex viscosity trends were found to depend on several experimental factors, namely: the screw profile, the clay nature and the possible presence of a compatibilizer. The most significant differences between the rheological curves were observed at low angular frequencies, for which the nanocomposites, according to the high degree of intercalation–exfoliation, showed the highest complex viscosity and G' values. Besides, using the power law expression ($\eta = A\omega^n$) it was possible to calculate the shear thinning exponents n for the nanocomposites extruded with the configuration for intensive mixing and high residence time in the low frequency region (i.e., from 0.1 and 1 rad/s): it is worthy to underline that while the presence of the clay alone did not affect the shear thinning behavior: this latter was significantly changed in the presence of a compatibilizer, capable for increasing the clay dispersion within the polyolefin.

Then, the rheology of polyethylene-montmorillonite (Cloisite $Na^+$) nanocomposites prepared by means of a water-assisted melt-intercalation process performed in a twin-screw extruder was thoroughly investigated, working in small amplitude oscillatory frequency sweep mode [127]. It was found that the changes in the rheological behavior strictly depend on the level of dispersion of the nanoclay within the polymer matrix. In particular, for well-intercalated systems, in the terminal region, a rise in complex viscosity and elastic modulus, as well as a decrease in phase angle values was observed: these findings pointed out the existence of solid-like or pseudosolid-like rheological behavior in the investigated systems.

## 4.2. Particles

When inorganic particles are incorporated into a thermoplastic matrix, the resulting material shows a complex rheological behavior, quite different from the rheology of unfilled homopolymers.

From an overall point of view, the addition of particles to a molten polymer usually increases the melt viscosity and decreases the melt elasticity. Different factors, comprising the volume fraction of the particles, their shape, size and size distribution, as well as the level of dispersion are well-known to affect the bulk rheological properties of particulate-filled polymer melts.

Some of the advances on the rheology of LDPE filled with different particles will be summarized in the following.

LDPE-Fly ash composites at different filler loadings (ranging from 4.2 to 28.6 vol.%) were prepared using a Two Roll Mixing Mill [128]. Fly ash is a by-product derived from the combustion of coal in thermal power plants; more specifically, it is a mixture of oxides rich in silica, $Fe_2O_3$ and alumina. As assessed by rheological tests, both shear stress and shear viscosity increased with increasing the filler loading. Besides, two regions of shear thinning were observed at 200°C for the prepared composites. The first normal stress difference was found to decrease with increasing the fly ash loading and the temperature. Conversely, this parameter remained almost unchanged at low shear.

Quite recently, Cobalt/Aluminum and Nickel/Aluminum layered double hydroxide (LDH) were intercalated with dodecylsulphate, laurate, stearate and palmitate and then compounded in LDPE at different loadings, ranging from 0.2 to 7.0 wt.% [129]. Low LDH loadings (i.e., below 2 wt.%) did not substantially affect LDPE shear moduli; conversely, with increasing the filler loading, the rheological curves showed a decrease of G' and G" moduli: this finding was ascribed to phase separation phenomena, occurring between LDH and LDPE regions. Besides, as revealed by the master curves of the unfilled polymer and its compounds with LDH, built using a vertical shift factor obtained at high frequency range (~100 $rad.s^{-1}$), no crossing between G' and G" was observed, and G" was always higher than G' throughout the frequency sweep: these findings confirmed the absence of transient networks in the molten state for the LDPE at the working temperature (190°C).

Liang investigated the effects of temperature, elongation strain rate, die extrusion velocity, and nano-ZnO loading on the rheological properties of LDPE-ZnO nanocomposites prepared through a melt spinning technique [130]: this allowed understanding the rheological behavior mechanisms occurring in the composites during the melt elongation flow. To this aim, different LDPE-ZnO compounds (filler concentration: 0.2 to 4 wt.%) were designed, extruded in a twin-screw extruder and analyzed in a constant rate type of capillary rheometer. It was found that the effect of the nano-ZnO loading and elongation strain rate on the rheological behavior of composite melts was irregular. In particular, the melt elongation viscosity decreased with increasing the temperature, approximately following the Arrhenius equation. Furthermore, the melt elongation viscosity of the prepared composites containing low amounts of ZnO (within 0.4 and 0.8 wt.%) was lower in the whole composite system and showed the "roller effect". In particular, at elongation strain rates below 0.3 $s^{-1}$, the melt elongation viscosity raised and then lowered with the increasing elongation strain rate; meanwhile, the melt elongation stress reached a maximum around an elongation strain rate of 0.3 $s^{-1}$. This finding was ascribed to stress hardening effects.

Lin and co-workers investigated the rheological behavior of LDPE-MgO nanocomposites at different filler loadings (from 0.25 to 20 wt.%), using a rheometer working in oscillatory mode [131]. Notwithstanding a typical shear thinning behavior observed for all the nanocomposite systems, the complex viscosity was further reduced for the compounds incorporating low ZnO amounts (i.e., up to 1 wt.% of nanofiller). This finding was ascribed to the increase of free volume between the macromolecular chains and the nanoparticles, i.e., to an interface effect.

Recently, the rheological properties of LDPE composites incorporating LDH modified with disodium 2,2'-((1,1'-biphenyl)-4,4'-diyldivinylene)bis (benzenesulphonate), employed as fluorescent whitening agent, i.e., as a chemical compound that absorbs light in the ultraviolet region and re-emits in the blue region, were thoroughly investigated [132]. In particular, it was found that G' and G" moduli showed constant values in a wide range of strain; besides, G" values were higher than G', and when the temperature was raised, a decrease of the elastic and viscous moduli was observed. Finally, as assessed in frequency sweep tests, both G' and G" were slightly affected by the incorporation of the modified LDH.

## 4.3. Carbon-Based Fillers

Several carbonaceous materials (such as carbon black, carbon nanotubes, graphene, fullerenes, among a few to mention) have been incorporated into different thermoplastic and thermosetting matrices, aiming to study the thermal, rheological, electrical, mechanical and barrier properties of the resulting materials. However, only few papers deal with the rheological behavior of LDPE-carbon based filler composites, which will be summarized in the following.

Gaska and co-workers prepared LDPE-based composites filled with different amounts of graphene nanoplatelets (ranging in between 1 and 7.5 wt.%) [133]. The rheological behavior of the obtained nanocomposites was investigated using a rheometer with plate-plate geometry; linear and nonlinear viscoelastic oscillatory shear tests were carried out, aiming at identifying the rheological percolation thresholds. As far as G' and G" dependence in the terminal region is considered, a slight increase with the increasing graphene nanoplatelets loading in the whole angular frequency range was observed; however, it was not possible to prove the existence of an additional elastic contribution in the lower limit of the angular frequency range. Finally, the rheological percolation was observed for the systems containing 7.5 wt.% of the nanofiller.

Finally, in a recent work, LDPE-based nanocomposites containing different amounts of graphene (namely, 0.5, 1 and 3 wt.%) were prepared by melt compounding, using a mini-lab twin-screw mixer [134]. The Newtonian behavior shown by LDPE at very low frequencies totally disappeared for the nanocomposites containing the highest graphene loading, which exhibited shear thinning behavior unchanged with respect to the unfilled matrix. Besides, the viscosity change from liquid-like to solid-like at low frequencies was attributed to the formation of interlocked network structures in the nanocomposites, ascribed to hindered motion of macromolecular chain segments.

## CONCLUSION

Though low density poly(ethylene) is a very well-established and cheap commodity, its processing is strictly related to the whole comprehension of the rheological behavior: therefore, the setup of such process parameters as temperature and shear rate is very important in order to obtain the most performing components or parts. For sure, the processing parameters depend on the architecture of the polyolefin, its molecular weight and molecular weight distribution, as well as on the branching degree. Therefore, a deep knowledge of all these parameters is really a key issue: rheological tests, performed according to different experimental conditions, can effectively help the designers and the manufacturers, leading to optimized conditions. Besides, as the use of neat LDPE homopolymers is somehow limited to specific applications (mainly including packaging and biomedical applications), a deep understanding of the effects of micro-to-nanofillers at different loadings on the polymer processability is undoubtedly of high importance, especially for high-tech advanced applications. In fact, the sensitivity of the rheological features with the structure and morphology of filled LDPE systems can help to discriminate the different morphologies that the standard characterization techniques are not able to elucidate.

## REFERENCES

[1] Osswald, Tim. & Rudolph, Natalie. (2015). *Polymer Rheology. Fundamentals and Applications*. Munich: Hanser Publishers.

[2] Gahleitner, M. (2001). Melt rheology of polyolefins. *Progress in Polymer Science, 26* (6), 895-944.
[3] Montgomery T. Shaw. (2012). *Introduction to Polymer Rheology.* Hoboken (NJ): John Wiley & Sons, Inc.
[4] Osswald, Tim A. & Menges, Georg. (2012). *Material Science of Polymers for Engineers (3rd Edition).* Munich: Hanser Publishers.
[5] Townsend, J. M., Beck, E. C., Gehrke, S. H., Berkland, C. J. & Detamore, M. S. (2019). Flow behavior prior to crosslinking: The need for precursor rheology for placement of hydrogels in medical applications and for 3D bioprinting. *Progress in Polymer Science, 91,* 126-140.
[6] Malkin, Alexander Y. & Isayev, Avraam. (2017). *Rheology - Concept, Methods, and Applications (3rd Edition).* Toronto: ChemTec Publishing.
[7] Kavanagh, G. M. & Ross-Murphy, S. B. (1998). Rheological characterization of polymer gels. *Progress in Polymer Science, 23* (3), 533-562.
[8] Dealy, John M. & Larson, Ronald G. (2006). *Structure and Rheology of Molten Polymers - From Structure to Flow Behavior and Back Again.* Munich: Hanser Publishers.
[9] Macosko, Christopher W. (1994). *Rheology: Principles, Measurements and Applications.* Hoboken (NJ): John Wiley & Sons, Inc.
[10] Savvas, T. A., Markatos, N. C. & Papaspyrides, C. D. (1994). On the flow of non-Newtonian polymer solutions. *Applied Mathematical Modelling, 18* (1), 14-22.
[11] Giles, Harold F. Jr, Wagner, John R. & Eldridge M. M. (2005). *Extrusion: The Definitive Processing Guide and Handbook.* Norwich (NY): William Andrew Inc.
[12] Denn, Morton M. (2008). *Polymer Melt Processing. Foundations in Fluid Mechanics and Heat Transfer.* New York (NY): Cambridge University Press.
[13] Münstedt, Helmut. (2019). *Elastic Behavior of Polymer Melts - Rheology and Processing.* Munich: Hanser Publishers.

[14] McCrum, N. G., Buckley, C. P. & Bucknall, C. B. (1997). "Viscoelasticity" in *Principles of Polymer Engineering (2nd Edition)*, edited by McCrum, N. G., Buckley, C. P. and Bucknall, C. B., 117-178. New York: Oxford University Press.

[15] Dealy, John M. & Wang, Jian. (2013). *Melt Rheology and its Applications in the Plastics Industry*. Dordrecht: Springer Netherlands.

[16] Dealy, John M. & Wissbrun, K. F. (1999). *Melt Rheology and Its Role in Plastics Processing. Theory and Applications*. Dordrecht: Springer Netherlands.

[17] Chung, Chan I. (2019). *Extrusion of Polymers - Theory and Practice (3rd Edition)*. Munich: Hanser Publishers.

[18] Cheremisinoff, Nicholas P. (2017). *Introduction to Polymer Rheology and Processing*. Boca Raton (FL): CRC Press.

[19] Han, Chang D. (2007). *Rheology and Processing of Polymeric Materials, Volume 2 - Polymer Processing*. New York: Oxford University Press.

[20] Collyer, A. A. & Utracki L. A. (1990). *Polymer Rheology and Processing*. New York (NY): Elsevier Science Publishers LTD.

[21] Han, Chang D. (2007). *Rheology and Processing of Polymeric Materials, Volume 1 - Polymer Rheology*. New York: Oxford University Press.

[22] Kontopoulou, Marianna. (2012). *Applied Polymer Rheology. Polymeric Fluids with Industrial Applications*. Hoboken (NJ): John Wiley & Sons, Inc.

[23] Sabu, T., Muller, R. & Abraham, J. (2016). *Rheology and Processing of Polymeric Nanocomposites*. Hoboken (NJ): John Wiley & Sons, Inc.

[24] Tadmor, Zehev & Gogos, Zehev. (2006). *Principles of Polymer Processing*. Hoboken (NJ): John Wiley & Sons, Inc.

[25] Fetters, L. J., Lohse, D. J., Richter, D., Witten, T. A. & Zirkel, A. (1994) Connection between Polymer Molecular Weight, Density, Chain Dimensions, and Melt Viscoelastic Properties. *Macromolecules, 27* (17), 4639-4647.

[26] Münstedt, H. & Schwarzl F. R. (2014). "Rheological Properties and Molecular Structure." In *Deformation and Flow of Polymeric Materials*, edited by Helmut Münstedt and Friedrich R. Schwarzl, 419-467. Dordrecht: Springer Netherlands.
[27] Santamaria, A. (1985). Influence of long chain branching in melt rheology and processing of low density polyethylene. *Materials Chemistry and Physics, 12* (1), 1-28.
[28] Yamaguchi, M. & Abe, S. (1999). LLDPE/LDPE Blends. I. Rheological, Thermal, and Mechanical Properties. *Journal of Applied Polymer Science, 74*, 3153-3159.
[29] Gahleitner, M. (2001). Melt rheology of polyolefins, *Progress in Polymer Science, 26* (6), 895-944.
[30] Rueda, M. M., Auscher, M. C., Fulchiron, R., Périé, T., Martin, G., Sonntag, P. & Cassagnau, P. (2017). Rheology and applications of highly filled polymers: A review of current understanding. *Progress in Polymer Science, 66*, 22-53.
[31] Leblanc, J. L. (2002). Rubber-filler interactions and rheological properties in filled compounds. *Progress in Polymer Science, 27* (4), 627-687.
[32] Sinha Ray, S. & Okamoto, M. (2003). Polymer/layered silicate nanocomposites: A review from preparation to processing, *Progress in Polymer Science, 28* (11), 1539-1641.
[33] Cassagnau, P. (2008). Melt rheology of organoclay and fumed silica nanocomposites. *Polymer, 49* (9), 2183-2196.
[34] Krishnamoorti, R. & Yurekli, K. (2001). Rheology of polymer layered silicate nanocomposites. *Current Opinion in Colloid and Interface Science, 6*(5-6), 464-470.
[35] Wood-Adams, P. M. & Dealy, J. M. (2000). Using Rheological Data to Determine the Branching Level in Metallocene Polyethylenes. *Macromolecules, 33*, 7481-7488.
[36] Larson, Ronald G. (1999). *The Structure and Rheology of Complex Fluids*. New Work: Oxford University Press.
[37] Dantzig, Jonathan A. & Tucker, Charles L. (2011). *Modeling in Materials Processing*. New York (NY): Cambridge University Press.

[38] Dawson P. C. (1999). "Flow Properties of Molten Polymers" in *Mechanical Properties and Testing of Polymers*, edited by Swallowe G.M. Dordrecht: Springer Netherlands.
[39] Münstedt, Helmut. (2016). *Rheological and Morphological Properties of Dispersed Polymeric Materials. Filled Polymers and Polymer Blends*. Munich: Hanser Publishers.
[40] Drabek, J., Zatloukal, M. & Martyn, M. (2018). Effect of molecular weight on secondary Newtonian plateau at high shear rates for linear isotactic melt blown polypropylenes, *Journal of Non-Newtonian Fluid Mechanics, 251*, 107-118.
[41] Wagner, M. H. & Meissner, J. (1980). Network disentanglement and time-dependent flow behaviour of polymer melts. *Macromolecular Chemistry and Physics, 181* (7), 1533-1550.
[42] Chatterjee, T. & Krishnamoorti, R. (2013). Rheology of Polymer-Carbon Nanotubes Composites. *Soft Matter, 9*, 9515-9529.
[43] Hieber, C. A. & Chiang, H. H. (1989). Some correlations involving the shear viscosity of polystyrene melts. *Rheologica Acta, 28*, 321-332.
[44] Hieber, C. A. & Chiang, H. H. (1992). Shear-rate-dependence modeling of polymer melt viscosity. *Polymer Engineering and Science, 32* (14), 931-938.
[45] Yu, Wei. (2013). "Rheological Measurements" in *Encyclopedia of Polymer Science and Technology*, edited by Krzysztof Matyjaszewski. New York (NY): John Wiley & Sons.
[46] Cheremisinoff, Nicholas P. (1996). *Polymer Characterization. Laboratory Techniques and Analysis*. Westwood (NJ): Noyes Publications.
[47] Bird, R. B., Armstrong, R. & Hassager, O. (1987). *Dynamics of Polymer Liquids, Vol. I: Fluid Mechanics, 2nd edn*. New York (NY): John Wiley & Sons.
[48] Lee, H. M. & Park, O. O. (1994). Rheology and dynamics of immiscible polymer blends. *Journal of Rheology, 38*(5), 1405-1425
[49] Papanicolaou, G. C. & Zaoutsos, S. P. "Viscoelastic constitutive modeling of creep and stress relaxation in polymers and polymer matrix composites" in *Creep and Fatigue in Polymer Matrix*

*Composites (Second Edition)*, edited by Rui Miranda Guedes, 3-59. Cambridge: Woodhead Publishing, Elsevier.

[50] Domínguez, J. C. (2018). "Rheology and curing process of thermosets" in *Thermosets (Second Edition)*, edited by Qipeng Guo, 115-146. Cambridge: Elsevier.

[51] Ferry, J. D. (1980). *Viscoelastic properties of polymers. 3rd edit.* New York (NY): John Wiley & Sons.

[52] Tschoegl, N. W. (1989). *The phenomenological theory of linear viscoelastic behavior: an introduction.* New York: Springer-Verlag.

[53] Giacomin, A. J. & Dealy, J. M. (1993). "Large-amplitude oscillatory shear" in *Techniques in rheological measurements*, edited by Collyer A. A., 99–121. London: Chapman & Hall.

[54] Vega, J. F., Santamaría, A., Muñoz-Escalona, A. & Lafuente P. (1998). Small-Amplitude Oscillatory Shear Flow Measurements as a Tool to Detect Very Low Amounts of Long Chain Branching in Polyethylenes. *Macromolecules, 31* (11), 3639-3647.

[55] Domenech, T., Zouari, R., Vergnes, B. & Peuvrel-Disdier, E. (2014). Formation of Fractal-like Structure in Organoclay-Based Polypropylene Nanocomposites. *Macromolecules, 47*, 3417-3427.

[56] Hyun, K., Wilhelm, M., Klein, C. O., Cho, K. S., Nam, J. G., Ahn, K. H., Lee, S. J., Ewoldt, R. H. & McKinley, G. H. (2011). A review of nonlinear oscillatory shear tests: Analysis and application of large amplitude oscillatory shear (LAOS), *Progress in Polymer Science, 36* (12), 1697-1753.

[57] Arrigo, R., Jagdale, P., Bartoli, M., Tagliaferro, A. & Malucelli, G. (2019). Structure–Property Relationships in Polyethylene-Based Composites Filled with Biochar Derived from Waste Coffee Grounds. *Polymers, 11*(8), 1336.

[58] La Mantia, F. P., Dintcheva, N. T., Filippone, G. & Acierno, D. (1996). Structure and Dynamics of Polyethylene/Clay Films. *Journal of Applied Polymer Science, 102*, 4749-4758.

[59] Osaki, K. (1993). On the damping function of shear relaxation modulus for entangled polymers. *Rheologica Acta, 32*, 429-437.

[60] Arrigo, R., Teresi, R., Gambarotti, C., Parisi, F., Lazzara, G. & Dintcheva, N. T. (2018). Sonication-Induced Modification of Carbon Nanotubes: Effect on the Rheological and Thermo-Oxidative Behaviour of Polymer-Based Nanocomposites. *Materials, 11*, 383.

[61] Münstedt, H. & Schwarzl F. R. (2014). "Extensional Rheology." In *Deformation and Flow of Polymeric Materials*, edited by Helmut Münstedt and Friedrich R. Schwarzl, 387-418. Dordrecht: Springer Netherlands.

[62] Okamoto, M., Kojima, A. & Kotaka, T. (1998). Elongational Flow Birefringence of Reactor-Made Linear Low-Density Polyethylene. *Macromolecules, 31*, 5158-5159.

[63] Bach, A., Rasmussen, H. K. & Hassager, O. (2003). Extensional viscosity for polymer melts measured in the filament stretching rheometer. *Journal of Rheology, 47*, 429.

[64] Hertel, D., Valette, R. & Munstedt, H. (2008). Three-dimensional entrance flow of a low-density polyethylene (LDPE) and a linear low-density polyethylene (LLDPE) into a slit die. *Journal of Non-Newtonian Fluid Mechanics, 153*, 82-94.

[65] Wagner M. H. (1999). "Elongational viscosity and its meaning for the praxis" in *Polypropylene. Polymer Science and Technology Series*, vol 2, edited by J. Karger-Kocsis J. Dordrecht: Springer Netherlands.

[66] Hsu, J. C. & Flumerfelt, R. W. (1975). Rheological Applications of a Drop Elongation Experiment. *Transactions of the Society of Rheology, 19*, 523.

[67] Sentmanat, M., Wang, B. N. & McKinley, G. H. (2005). Measuring the transient extensional rheology of polyethylene melts using the SER universal testing platform. *Journal of Rheology, 49*, 585-606.

[68] Stange, J. & Münstedt, H. (2006). Rheological properties and foaming behavior of polypropylenes with different molecular structures. *Journal of Rheology, 50*(6), 907-923.

[69] La Mantia, F. P., Valenza, A. & Acienro, D. (1986). Influence of Long Chain Branching on the Elongational Behaviour of Different Polyethylenes and Their Blends. *Polymer Bulletin, 15*, 381-387.

[70] La Mantia, F. P., Arrigo, R. & Morreale, M. (2014). Effect of the orientation and rheological behaviour of biodegradable polymer nanocomposites. *European Polymer Journal, 54*, 11-17.

[71] Guadarrama-Medina, T. d. J., Pérez-González, J. & de Vargas, L. (2005). Enhanced melt strength and stretching of linear low-density polyethylene extruded under strong slip conditions. *Rheologica Acta, 44*, 278-286.

[72] Acierno, D., Brancaccio, A., Curto, D., La Mantia, F. P. & Valenza, A. (1985). Molecular Weight Dependency of Rheological Characteristics of Linear Low Density Polyethylene. *Journal of Rheology, 29*, 323-334.

[73] Mendelson, R. A., Bowles, W. A. & Finger, F. L. (1970). Effect of molecular structure on polyethylene melt rheology. I. Low-shear behavior. *Journal of Polymer Science: Part A-2: Polymer Physics, 8*, 105-126.

[74] Gabriel, C. & Lilge, D. (2006). Molecular mass dependence of the zero shear-rate viscosity of LDPE melts: evidence of an exponential behaviour. *Rheologica Acta, 45*, 995.

[75] Munstedt, H. (2011). Rheological properties and molecular structure of polymer melts. *Soft Matter, 7*, 2273-2283.

[76] Onogi, S., Masuda, T. & Kitagawa, K. (1970). Rheological Properties of Anionic Polystyrenes. I. Dynamic Viscoelasticity of Narrow-Distribution Polystyrenes. *Macromolecules, 3*(2), 109-116.

[77] Pechhold, W., Soden, W. & Stoll, B. (1981). Shear compliance of polymer melts and its dependence on molecular weight. *Die Makromolekulare Chemie, 182* (2), 573-581.

[78] Acierno, D., La Mantia, F. P., Romanini, D. & Savadori, A. (1985). An experimental investigation of the shear behaviour of polyethylenes with different structures. *Rheologica Acta, 24*, 566-573.

[79] Stadler, F. J., Piel, C., Kaschta, J., Rulhoff, S., Kaminsky, W. & Münstedt, H. (2006). Dependence of the zero shear-rate viscosity and the viscosity function of linear high-density polyethylenes on the mass-average molar mass and polydispersity. *Rheologica Acta, 45* (5), 755-764.

[80] Auhl, D., Stange, J., Münstedt, H., Krause, B., Voigt, D., Lederer, A., Lappan, U. & Lunkwitz, K. (2004). Long-Chain Branched Polypropylenes by Electron Beam Irradiation and Their Rheological Properties. *Macromolecules, 37* (25), 9465-9472.

[81] Wood-Adams, P. M., Dealy, J. M., Degroot, A. W. & Redwine, O. R. (2000). Effect of molecular structure on the linear viscoelastic behavior of polyethylene. *Macromolecules, 33* (20), 7489-7499.

[82] Raju, V. R., Smith, G. G., Marin, G., Knox, J. R. & Graessley, W. W. (1979). Properties of amorphous and crystallizable hydrocarbon polymers. I. Melt rheology of fractions of linear polyethylene. *Journal of Polymer Science: Polymer Physics Edition, 17* (7), 1183-1195.

[83] Nichetti, D. & Manas-Zloczowera, I. (1998). Viscosity model for polydisperse polymer melts. *Journal of Rheology, 42*(4), 951-969.

[84] Liu, C., Wang, J. & He, J. (2002). Rheological and thermal properties of m-LLDPE blends with m-HDPE and LDPE. *Polymer, 43* (13), 3811-3818.

[85] Han, C. D. & Villamizar, C. A. (1978). Effects of molecular weight distribution and long-chain branching on the viscoelastic properties of high- and low-density polyethylene melts. *Journal of Applied Polymer Science, 22* (6), 1677-1700.

[86] La Mantia, F. P., Valenza, A. & Acierno, D. (1983). A comprehensive experimental study of the rheological behaviour of polyethylene. II. Die-swell and normal stresses. *Rheologica Acta, 22*, 308-312.

[87] Koopmans, R. J. & Molenaar, J. (2004). The "Sharkskin Effect" in polymer extrusion. *Polymer Engineering and Science, 38*(1), 101-107.

[88] Kissi, N. E., Piau, J. M. & Toussaint, F. (1997). Sharkskin and cracking of polymer melt extrudates. *Journal of Non-Newtonian Fluid Mechanics, 68* (2-3), 271-290.

[89] Venet, C. & Vergnes, B. (1997). Experimental characterization of sharkskin in polyethylenes. *Journal of Rheology, 41* (1), 873-892.

[90] Burghelea, T. I., Griess, H. J. & Münstedt, H. (2010). Comparative investigations of surface instabilities ("sharkskin") of a linear and a long-chain branched polyethylene. *Journal of Non-Newtonian Fluid Mechanics, 165* (19-20), 1093-1104.

[91] Allal, A., Lavernhe, A., Vergnes, B. & Marin, G. (2006). Relationships between molecular structure and sharkskin defect for linear polymers. *Journal of Non-Newtonian Fluid Mechanics*, *134* (1-3), 127-135.

[92] Yamaguchi, M., Miyata, H., Tan, V. & Gogos, C. G. (2002). Relation between molecular structure and flow instability for ethylene/α-olefin copolymers. *Polymer*, *43* (19), 5249-5255.

[93] Ansari, M., Derakhshandeh, M., Doufas, A. A., Tomkovic, T. & Hatzikiriakos, S. G. (2018). The role of microstructure on melt fracture of linear low density polyethylenes. *Polymer Testing*, *67*, 266-274.

[94] Wood-Adams, P. M. & Dealy, J. M. (1996). Use of rheological measurements to estimate the molecular weight distribution of linear polyethylene. *Journal of Rheology*, *40* (5), 761-778.

[95] Liu, Y. & Shaw, M. T. (1998). Obtaining molecular-weight distribution information from the viscosity data of linear polymer melts. *Journal of Rheology*, *42*, 453.

[96] Bersted, B. H. (1975). An empirical model relating the molecular weight distribution of high-density to the shear dependence of the steady shear viscosity. *Journal of Applied Polymer Science*, *19*, 2167-2177.

[97] Wu, S. (1985). Polymer molecular-weight distribution from dynamic melt viscoelasticity. *Polymer Engineering and Science*, *25*, 122-128.

[98] Bersted, B. H., Slee, J. D. & Richter, C. A. (1981). Prediction of rheological behavior of branched polyethylene from molecular structure. *Journal of Applied Polymer Science*, *26* (3), 1001-1014.

[99] Santamaria, A. (1985). Influence of Long Chain Branching in Melt Rheology and Processing Of Low Density Polyethylene. *Materials Chemistry and Physics*, *12*, 1-28.

[100] Dealy, John M. & Wang, Jian. (2013). "Rheology and Molecular Structure" in *Melt Rheology and its Applications in the Plastics Industry*, edited by Dealy, John M. and Wang, Jian, 181-200. Dordrecht: Springer Netherlands.

[101] Gotsis, A. D. (2012). "Branched Polyolefins" in *Applied Polymer Rheology. Polymeric Fluids with Industrial Applications*, edited by Kontopoulou, Marianna. Hoboken (NJ): John Wiley & Sons, Inc.

[102] Zhu, X., Zhou, Y. & Yan, D. (2011). Influence of Branching Architecture on Polymer Properties. *Journal of Polymer Science Part B: Polymer Physics, 49*, 1277-1286.
[103] Fetters, L. J., Lohse, D. J., Richter, D., Witten, T. A. & Zirkel, A. (1994). Connection between Polymer Molecular Weight, Density, Chain Dimensions, and Melt Viscoelastic Properties. *Macromolecules, 27* (17), 4639-4647.
[104] Stadler, F. J., Piel, C., Kaminsky, W. & Munstedt, H. (2006). Rheological Characterization of Long-chain Branched Polyethylenes and Comparison with Classical Analytical Methods. *Macromolecular Symposia, 236*, 209-218.
[105] Keßner, U., Kaschta, J. & Münstedt, H. (2009). Determination of method-invariant activation energies of long-chain branched low-density polyethylenes. *Journal of Rheology, 53*, 1001-1016.
[106] Read, D. J. & McLeish, T. C. B. (2001). Molecular Rheology and Statistics of Long Chain Branched Metallocene-Catalyzed Polyolefins. *Macromolecules, 34*, 1928-1945.
[107] Bersted, B. H. (1985). On the Effects of Very Low Levels of Long Chain Branching on Rheological Behavior in Polyethylene. *Journal of Applied Polymer Science, 30*, 3751-3765.
[108] Yan, D., Wang, W. J. & Zhu, S. (1999). Effect of long chain branching on rheological properties of metallocene polyethylene. *Polymer, 40* (7), 1737-1744.
[109] Stadler, F. J. & Muunstedt, H. (2009). Correlations between the Shape of Viscosity Functions and the Molecular Structure of Long-Chain Branched Polyethylenes. *Macromolecular Materials and Engineering, 294*, 25-34.
[110] La Mantia, F. P., Scaffaro, R., Carianni, G. & Mariani, P. (2005). Rheological Properties of Different Film Blowing Polyethylene Samples Under Shear and Elongational Flow. *Macromolecular Materials and Engineering, 290*, 159-164.
[111] Valenza, A., La Mantia, F. P. & Acierno, D. (1988). Shear and Elongational Rheology of Linear Low Density Polyethylenes with Different Structures. *European Polymer Journal, 24*(1), 81-85.

[112] Stadler, F. J., Kaschta, J. & Munstedt, H. (2008). Thermorheological Behavior of Various Long-Chain Branched Polyethylenes. *Macromolecules, 41*, 1328-1333.

[113] Laun, H. M. & Münstedt, H. (1978). Elongational behaviour of a low density polyethylene melt I. Strain rate and stress dependence of viscosity and recoverable strain in the steady-state. Comparison with shear data. Influence of interfacial tension. *Rheologica Acta, 17*, 415-425.

[114] Wolff, F., Resch, J. A., Kaschta, J. & Münstedt, H. (2010). Comparison of viscous and elastic properties of polyolefin melts in shear and elongation. *Rheologica Acta, 49*, 95-103.

[115] Munstedt, H., Steffl, T. & Malmberg, A. (2005). Correlation between rheological behaviour in uniaxial elongation and film blowing properties of various polyethylenes. *Rheologica Acta, 45*, 14-22.

[116] Münstedt, H. & Laun, H. M. (1981). Elongational properties and molecular structure of polyethylene melts. *Rheologica Acta, 20* (3), 211-221.

[117] Stadler, F. J., Kaschta, J., Münstedt, H., Becker, F. & Buback, M. (2009). Influence of molar mass distribution and long-chain branching on strain hardening of low density polyethylene. *Rheologica Acta, 48*, 479-490.

[118] La Mantia, F. P. & Acierno, D. (1985). Influence of the Molecular Structure on the Melt Strength and Extensibility of Polyethylenes. *Polymer Engineering and Science, 25*(5), 279-283.

[119] Sinha Ray, S. & Okamoto, M. (2003). Polymer/layered silicate nanocomposites: a review from preparation to processing. *Progress in Polymer Science, 28*, 1539–1641.

[120] Devalckenaere, M., Je, R., Dubois, P. & Kubies, D. (2002). Poly(E-carprolactone)/clay nanocomposites prepared by melt intercalation: mechanical, thermal and rheological properties. *Polymer, 43*(14), 4017-4023.

[121] Hoffmann, B., Dietrich, C., Thomann, R., Friedrich, C. & Mu, R. (2000). Morphology and rheology of polystyrene nanocomposites

based upon organoclay. *Macromolecular Rapid Communications, 61*(1), 57-61.
[122] U, R. K. & Yurekli, K. (2001). Rheology of polymer layered silicate nanocomposites. *Current Opinion in Colloid Interface Science, 6*(5-6), 464-470.
[123] Xu, L., Reeder, S., Thopasridharan, M., Ren, J., Shipp, D. A. & Krishnamoorti, R. (2005). Structure and melt rheology of polystyrene-based layered silicate nanocomposites. *Nanotechnology, 16*(7), S514-S521.
[124] Wagener, R. & Reisinger, T. J. G. (2003). A rheological method to compare the degree of exfoliation of nanocomposites. *Polymer, 44*(24), 7513-7518.
[125] Malucelli, G., Ronchetti, S., Lak, N., Priola, A., Tzankova Dintcheva, N. & La Mantia, F. P. (2007). Intercalation effects in LDPE/o-montmorillonites nanocomposites. *European Polymer Journal, 43*, 328–335.
[126] Villanueva, M. P., Cabedo, L., Gimenez, E., Lagaron, J. M., Coates, P. D. & Kelly A. L. (2009). Study of the dispersion of nanoclays in a LDPE matrix using microscopy and in-process ultrasonic monitoring. *Polymer Testing, 28*, 277–287.
[127] Shahabadi, S. I. S. & Garmabi, H. (2012). Polyethylene/$Na^+$-montmorillonite composites prepared by slurry-fed melt intercalation: Response surface analysis of rheological behavior. *Journal of Reinforced Plastics and Composites, 31*(14), 977–988.
[128] Hashmi, S. A. R., Sharma, P. & Chand, N. (2008). Thermal and Rheological Behavior of Ultrafine Fly Ash Filled LDPE Composites. *Journal of Applied Polymer Science, 107*, 2196–2202.
[129] Jaerger, S., Leuteritz, A., Alves de Freitas, R. & Wypych, F. Rheological properties of low-density polyethylene filled with hydrophobic Co(Ni)-Al layered double hydroxides. *Polímeros, 29*(1), e2019007, 2019.
[130] Ji Zhao Liang. Melt Elongation Flow Behavior of Low-Density Polyethylene Composites Filled with Nanoscale Zinc Oxide. *Journal*

of Testing and Evaluation https://doi.org/10.1520/JTE20180301. ISSN 0090-3973.

[131] Lin, X., Wu, Y. H., Tang, L. Y., Yang, M. H., Ren, D. Y., Zha, J. W. & Dang, Z. M. (2016). Experimental study of the rheological, mechanical, and dielectric properties of MgO/LDPE nanocomposites. *J. Appl. Polym. Sci.*, *133*(7), 43038.

[132] Monzó, F., Caparró, A. V., Pérez-Pérez, D., Arribas, A. & Pamies, R. (2019). Synthesis and Characterization of New Layered Double Hydroxide-Polyolefin Film Nanocomposites with Special Optical Properties. *Materials*, *12*, 3580.

[133] Gaska, K., Kádár, R., Rybak, A., Siwek, A. & Gubanski, S. (2017). Gas Barrier, Thermal, Mechanical and Rheological Properties of Highly Aligned Graphene-LDPE Nanocomposites, *Polymers*, *9*, 294.

[134] Sabet, M. & Soleimani, H. (2019). Inclusion of graphene on low-density polyethylene composite properties. *International Journal of Plastics Technology*, *23*, 218–228.

## BIOGRAPHICAL SKETCHES

### *Rossella Arrigo*

**Affiliation:** Politecnico di Torino, Dept. of Applied Science and Technology, Viale Teresa Michel 5, 15121 Alessandria, Italy

**Date of Birth:** 5 July 1983

**Education:** MSci in Chemical Engineering (2009), PhD in Chemical and Material Engineering (2014)

**Research and Professional Experience:** Assistant Professor of Materials Science and Technology

**Publications - Last Four Years (on peer-reviewed journals):**

Arrigo, Rossella., Antonioli, Diego., Lazzari, Massimo., Gianotti, Valentina., Laus, Michele., Montanaro, Laura. & Malucelli, Giulio (2018). Relaxation Dynamics in Polyethylene Glycol/Modified Hydrotalcite Nanocomposites, *Polymers, 10* (11), 1182.

Arrigo, Rossella., Dintcheva, Nadka., Guenzi, Monica. & Gambarotti, Cristian. (2016). Nano-hybrids based on Quercetin and Carbon Nanotubes with excellent antioxidant Activity, *Materials Letters, 180*, 7-10.

Arrigo, Rossella., Dintcheva, Nadka., Pampalone, Vito., Morici, Elisa., Guenzi, Monica. & Gambarotti, Cristian. (2016). Advanced nano-hybrids for thermo-oxidative resistant nanocomposites, *Journal of Materials Science, 51* (14), 6955-6966

Arrigo, Rossella., Dintcheva, Nadka., Tarantino, Giuseppe., Passaglia, Elisa., Coiai, Serena., Cicogna, Francesca., Filippi, Sara., Nasillo, Giorgio. &Chillura Martino, Delia. (2018). An insight into the interaction between functionalized thermoplastic elastomer and layered double hydroxides through rheological investigations, *Composites Part B: Engineering, 139*, 47-54.

Arrigo, Rossella., Jagdale, Pravin., Bartoli, Mattia., Tagliaferro, Alberto. & Malucelli, Giulio (2019). Structure–Property Relationships in Polyethylene-Based Composites Filled with Biochar Derived from Waste Coffee Grounds, *Polymers, 11*(8), 1336.

Arrigo, Rossella., Mascia, Leno., Clarke, Jane. & Malucelli, Giulio. (2020). Structure evolution of epoxidized natural rubber (ENR) in the melt state by time-resolved mechanical spectroscopy, *Materials, 13*(4), 946.

Arrigo, Rossella., Morici, Elisa. & Dintcheva, Nadka. (2017). High Performance Thermoplastic Elastomers/Carbon Nanotubes Nanocomposites: mechanical behaviour, rheology and durability, *Polymer Composites, 38*, E381–E391.

Arrigo, Rossella., Morici, Elisa., Cammarata, Marcello. & Dintcheva, Nadka. (2017). Rheological percolation threshold in high viscosity polymer/CNTs nanocomposites, *Journal of Engineering Mechanics, 143* (5), D4016006-1.

Arrigo, Rossella., Ronchetti, Silvia., Montanaro, Laura. & Malucelli, Giulio. (2018). Effects of the nanofiller size and aspect ratio on the thermal and rheological behavior of PEG nanocomposites containing boehmites or hydrotalcites, *Journal of Thermal Analysis and Calorimetry, 134,* 1167-1180.

Arrigo, Rossella., Teresi, Rosalia. & Dintcheva, Nadka. (2018). Mechanical and rheological properties of Polystyrene-block-Polybutadiene-block-Polystyrene copolymer reinforced with carbon nanotubes: effect of processing conditions, *Journal of Polymer Engineering, 38,* 107-117.

Arrigo, Rossella., Teresi, Rosalia., Gambarotti, Cristian., Parisi, Filippo., Lazzara, Giuseppe. & Dintcheva, Nadka. (2018). Sonication-induced modification of Carbon Nanotubes: effect on the rheological and thermo-oxidative behaviour of polymer-based nanocomposites, *Materials, 11* (3), 383.

D'Anna, Alessandra., Arrigo, Rossella. & Frache, Alberto. (2019). PLA/PHB blends: bio-compatibilizer effects, *Polymers, 11*(9), 1416.

Dintcheva, Nadka., Al-Malaika, Sahar., Arrigo, Rossella. & Morici, Elisa. (2017). Novel strategic approach for the thermo- and photo- oxidative stabilization of polyolefin/clay nanocomposites, *Polymer Degradation and Stability, 145,* 41-51.

Dintcheva, Nadka., Al-Malaika, Sahar., Morici, Elisa. & Arrigo, Rossella. (2017). Thermo-oxidative stabilization of polylactic acid-based nanocomposites through the incorporation of clay with in-built antioxidant activity, *Journal of Applied Polymer Science, 134* (24), 44974.

Dintcheva, Nadka., Arrigo, Rossella., Baiamonte, Marilena., Rizzarelli, Paola., Curcuruto, Giusy (2017) Concentration-dependent anti-/pro-oxidant activity of natural phenolic compounds in bio-polyesters, *Polymer Degradation and Stability, 142,* 21-28.

Dintcheva, Nadka., Arrigo, Rossella., Carroccio, Sabrina., Curcuruto, Giusy., Guenzi, Monica., Gambarotti, Cristian. & Filippone, Giovanni. (2016). Multi-functional Polyhedral Oligomeric Silsesquioxanefunctionalized Carbon Nanotubes for photo-oxidative stable Ultra-High Molecular Weight Polyethylene-based Nanocomposites, *European Polymer Journal, 75*, 525–537.

Dintcheva, Nadka., Arrigo, Rossella., Teresi, Rosalia. & Gambarotti, Cristian. (2017). Silanol-POSS as dispersing agents for Carbon Nanotubes in Polyamide, *Polymer Engineering and Science, 57* (6), 588-594.

Dintcheva, Nadka., Arrigo, Rossella., Teresi, Rosalia., Megna, Bartolomeo., Gambarotti, Cristian., Marullo, Salvatore. & D'Anna, Francesca. (2016). Tunable Radical Scavenging Activity of Carbon Nanotubes through Sonication, *Carbon, 107*, 240-247.

Dintcheva, Nadka., Catalano, Giulia., Arrigo, Rossella., Morici, Elisa., Cavallaro, Giuseppe., Lazzara, Giuseppe. & Bruno, Maurizio. (2016). Pluronic nanoparticles as anti-oxidant carriers for polymers, *Polymer Degradation and Stability, 134*, 194-201.

Dintcheva, Nadka., Filippone, Giovanni., Arrigo, Rossella. & La Mantia, Francesco Paolo. (2017). Effect of The Morphology of Clay-Containing Low Density Polyethylene/Polyamide Blends on Their Photo-Oxidation Resistance, *Journal of Nanomaterials*, 3549475.

Morici, Elisa., Di Bartolo, Alberto., Arrigo, Rossella. & Dintcheva, Nadka. (2016). Double bonds functionalized POSS: dispersion and crosslinking in polyethylene based hybrid obtained by reactive processing, *Polymer Bulletin, 73*, 3385-3400.

Morici, Elisa., Di Bartolo, Alberto., Arrigo, Rossella., Dintcheva, Nadka (2018) POSS Grafting on Polyethylene and Maleic Anhydride-Grafted Polyethylene by One-Step Reactive Melt Mixing, *Advances in Polymer Technology, 37* (2), 349-357.

Nasillo, Giorgio., Arrigo, Rossella., Dintcheva, Nadka., Morici, Elisa., Chillura Martino, Delia., Caponetti, Eugenio (2018) Polyamide-based fibers containing microwave exfoliated graphite nanoplatelets, *Advances in Polymer Technology, 37* (3), 786-797.

Rajczak, Ewa., Arrigo, Rossella. & Malucelli, Giulio. (2020). Thermal stability and flame retardance of EVA containing DNA-modified clays, *Thermochimica Acta*, vol. *686*, p. 178546.

Rizzo, Carla., Arrigo, Rossella., D'Anna, Francesca., Di Blasi, Francesco., Dintcheva, Nadka., Lazzara, Giuseppe., Parisi, Filippo., Riela, Serena., Spinelli, Giuseppe., Massaro, Marina (2017) Hybrid supramolecular gels of Fmoc-F/halloysite nanotubes: systems for sustained release of camptothecin. *Journal of Materials Chemistry B*, *5*, 3217-3229.

Rizzo, Carla., Arrigo, Rossella., Dintcheva, Nadka., Gallo, Giuseppe., Giannici, Francesco., Noto, Renato., Sutera, Alberto., Vitale, Paola. & D'Anna, Francesca. (2017). Supramolecular Hydro- and Ionogels: a study of their properties and antibacterial activity, *Chemistry – A European Journal*, *23*, 16297-16311.

## Giulio Malucelli

**Affiliation:** Politecnico di Torino, Dept. of Applied Science and Technology, Viale Teresa Michel 5, 15121 Alessandria, Italy

**Date of Birth:** 6 November 1967

**Education:** MSci in Chemical Engineering (1992), PhD in Chemistry (1996)

**Research and Professional Experience:** Full Professor of Materials Science and Technology

**Professional Appointments:**

- Founder of the Italian Macromolecular Association and Member (from 2011 to 2016) of the Steering Committee of the Italian Macromolecular Association
- Member (from 2013 to 2016) of the Steering Committee "Sustainable flame retardancy for textiles and related materials based on nanoparticles substituting conventional chemicals", FLARETEX (MP1105)

**Publications - Last Four Years (on peer-reviewed journals):**

Angela Castellano, Claudio Colleoni, Giuseppina Iacono, Alessio Mezzi, Maria Rosaria Plutino, Giulio Malucelli, Giuseppe Rosace (2019). Synthesis and characterization of a phosphorous/nitrogen based sol-gel coating as a novel halogen- and formaldehyde-free flame retardant finishing for cotton fabric. *Polymer Degradation and Stability*, vol. *162*, p. 148-159.

Arrigo, Rossella., Mascia, Leno., Clarke, Jane. & Malucelli, Giulio. (2020). Structure evolution of epoxidized natural rubber (ENR) in the melt state by time-resolved mechanical spectroscopy, *Materials*, *13*(4), 946.

Bosco, Francesca., Casale, Annalisa., Gribaudo, Giorgio., Mollea, Chiara. & Malucelli, Giulio. (2017). Nucleic acids from agro-industrial wastes: a green recovery method for fire retardant applications. In: *Industrial Crops And Products*, vol. *108*, pp. 208-218. - ISSN 0926-6690.

Branda, Francesco., Malucelli, Giulio., Durante, Massimo., Piccolo, Alessandro., Mazzei, Pierluigi., Costantini, Aniello., Silvestri, Brigida., Pennetta, Miriam, & Bifulco, Aurelio. (2016). Silica Treatments: A Fire Retardant Strategy for Hemp Fabric/Epoxy Composites. In: *Polymers*, vol. *8*, n. 313, pp. 1-17. - ISSN 2073-4360.

Casale, Annalisa., Bosco, Francesca., Malucelli, Giulio., Mollea, Chiara. & Periolatto, Monica. (2016). DNA-chitosan cross-linking and photografting to cotton fabrics to improve washing fastness of the fire-resistant finishing. In: *Cellulose*, vol. *23*, n. 6, pp. 3963-3984. - ISSN 0969-0239.

Dos Santos, R., Oberrauch, E., Banfi, M., Floris, M., Tognola, G., Marchizza, M., Malucelli, G. & Castrovinci, A. (2016). An online acquisition method for monitoring the surface growth of flame retardant protective layers. In: *Fire and Materials*, vol. *40*, pp. 544-553. - ISSN 1099-1018

Duraccio, D., Strongone, V., Malucelli, G., Auriemma, F., De Rosa, C., Mussano, F. D., Genova, T. & Faga, M. G. (2019). The role of alumina-zirconia loading on the mechanical and biological properties of

UHMWPE for biomedical applications. *Composites. Part B, Engineering*, vol. *164*, p. 800-808.

Ewa Rajczak, Bartosz Tylkowski, Magda Constantí, Monika Haponska, Boryana Trusheva, Giulio Malucelli, Marta Giamberini (2020). Preparation and characterization of UV-curable acrylic membranes embedding natural antioxidants, *Polymers*, *12*(2), 358 p. 1-16.

Ewa, Rajczak, Rossella Arrigo, Giulio Malucelli. (2020). Thermal stability and flame retardance of EVA containing DNA-modified clays, *Thermochimica Acta*, vol. *686*, p. 178546.

Giulio Malucelli, Marco Barbalini (2018). UV-curable acrylic coatings containing biomacromolecules: A new fire retardant strategy for ethylene-vinyl acetate copolymers. *Progress in Organic Coatings*, vol. *127*, p. 330-337.

Giulio, Malucelli (2018). Sol-Gel and Layer by Layer Methods for Conferring Multifunctional Features to Cellulosic Fabrics: An Overview. In: *Textiles: Advances in Research and Applications*/Boris Mahltig. Nova Science Publishers, Inc., Hauppauge, NY (USA), pp. 61-86. ISBN 978-1-53612-855-0.

Giulio, Malucelli. (2019). Textile finishing with biomacromolecules: A low environmental impact approach in flame retardancy. In: *Shahid ul-Islam. The Impact and Prospects of Green Chemistry for Textile Technology.* p. 251-279, Amsterdam: Elsevier, ISBN: 9780081024911, doi: 10.1016/B978-0-08-102491-1.00009-5.

Jelena Vasiljevic, Marija Colovic, Ivan Jerman, Barbara Simoncic, Andrej Demsar, Younes Samaki, Matic Sobak, Ervin Sest, Barbara Golja, Mirjam Leskovsek, Vili Bukosek, Jozef Medved, Marco Barbalini, Giulio Malucelli, Silvester Bolka (2019). *In situ* prepared polyamide 6/DOPO-derivative nanocomposite for melt-spinning of flame retardant textile filaments. *Polymer Degradation and Stability*, vol. *166*, p. 50-59.

Malucelli, G. & Rosace, G. (2019). Advances in Phosphorus-Based Flame Retardant Cotton Fabrics. In: *Fabien Salaün*. p. 107-165, Hauppauge, NY (USA): Nova Science Publishers, Inc., ISBN: 978-1-53615-006-3.

Malucelli, G. (2016). Hybrid organic/inorganic coatings through dual-cure processes: state of the art and perspectives. In: *Coatings*, vol. 6, n. 1, pp. 1-11. – ISSN.

Malucelli, Giulio. & Vaughan, Alun S. (2016). Degradation of Polymeric Micro- and Nanocomposites. In: *Tailoring of Nanocomposite Dielectrics: From Fundamentals to Devices and Applications*/Toshikatsu Tanaka and Alun S. Vaughan. Pan Stanford Publishing Pte. Ltd, Singapore, pp. 312-344. ISBN 978-981-4669-80-1.

Malucelli, Giulio. (2016). Biomacromolecules as Effective Green Flame Retardants for Textiles: An Overview. In: *Advances in Environmental Research*/Justin A. Daniels. Nova Science Publishers, Hauppauge, pp. 153-174. ISBN 978-1-63483-758-3.

Malucelli, Giulio. (2016). Layer-by-Layer nanostructured assemblies for the fire protection of fabrics. In: *Materials Letters*, vol. 166, pp. 339-342. - ISSN 0167-577X.

Malucelli, Giulio. (2016). Surface engineered fire protective coatings for fabrics through sol-gel and layer-by-layer methods: an overview. In: *Coatings*, vol. 6, n. 33, pp. 1-23. - ISSN 2079-6412.

Malucelli, Giulio. (2017). Graphene-based polymer nanocomposites: recent advances and still open challenges. In: *Current Graphene Science*, vol. *1*, pp. 16-25. - ISSN 2452-2732.

Malucelli, Giulio. (2017). High barrier composite materials based on renewable sources for food packaging applications. In: *Nanotechnology in Food Industry* (Volume *I-X*)/Alexandru Mihai Grumezescu. Alexandru Mihai Grumezescu, Amsterdam, pp. 45-78. ISBN 978-0-12-804302-8.

Malucelli, Giulio. (2017). Recent Advances in Uv-Curable Functional Coatings. In: *Advances in Materials Science Research*/Maryann C. Wythers. Nova Science Publishers, Inc., Hauppauge, NY, pp. 1-29. ISBN 978-1-53611-790-5.

Malucelli, Giulio. (2017). Synthesis and characterization of UV-LED curable nanocomposite coatings. In: *Current Organic Chemistry*, vol. *21*, pp. 1-8. - ISSN 1385-2728.

Malucelli, Giulio. (2020). Layer-by-Layer: an effective surface engineered strategy for flame retardant textiles, In: *Layer-By-Layer Deposition Development and Applications*, Nova Science Publishers, Inc., p. 183-215, ISBN: 978-1-53617-087-0.

Malucelli, Giulio., Fioravanti, Ambra., Francioso, Luca., De Pascali, Chiara., Signore, Maria Assunta., Carotta, Maria Cristina., Bonanno, Antonio. & Duraccio, Donatella. (2017). Preparation and characterization of UV-cured composite films containing ZnO nanostructures: effect of filler geometric features on piezoelectric response. In: *Progress in Organic Coatings*, vol. 109, pp. 45-54. - ISSN 0300-9440.

Mariani, A., Nuvoli, L., Sanna, D., Alzari, V., Nuvoli, D., Rassu, M.. & Malucelli, Giulio. (2018). Semi-interpenetrating polymer networks based on crosslinked poly(N-isopropyl acrylamide) and methylcellulose prepared by frontal polymerization. In: *Journal of Polymer Science. Part A, Polymer Chemistry*, vol. *56*, pp. 437-443. - ISSN 0887-624X.

Ortelli, Simona, Malucelli, Giulio, Blosi, Magda, Zanoni, Ilaria, Costa, Anna Luisa (2019). NanoTiO$_2$@DNA complex: a novel eco, durable, fire retardant design strategy for cotton textiles. *Journal of Colloid and Interface Science*, vol. *546*, p. 174-183.

Rassu, Mariella., Alzari, Valeria., Nuvoli, Daniele., Nuvoli, Luca., Sanna, Davide., Sanna, Vanna., Malucelli, Giulio. & Mariani, Alberto. (2017). Semi-interpenetrating polymer networks of methyl cellulose and polyacrylamide prepared by frontal polymerization. In: *Journal of Polymer Science. Part A, Polymer Chemistry*, vol. *55*, pp. 1268-1274. - ISSN 1099-0518.

Rosace, Giuseppe., Colleoni, Claudio., Guido, Emanuela. & Malucelli, Giulio. (2017). Phosphorus-Silica Sol-Gel Hybrid Coatings for Flame Retardant Cotton Fabrics. In: *Tekstilec*, vol. *60*, n. 1, pp. 29-35. - ISSN 2350-3696.

Rosace, Giuseppe., Colleoni, Claudio., Trovato, Valentina., Iacono, Giuseppina. & Malucelli, Giulio. (2017). Vinylphosphonic acid/methacrylamide system as a durable intumescent flame retardant for cotton fabric. In: *Cellulose*, vol. *24*, pp. 3095-3108. - ISSN 0969-0239.

Rossella Arrigo, Pravin Jagdale, Mattia Bartoli, Alberto Tagliaferro, Giulio Malucelli (2019). Structure–Property Relationships in Polyethylene-Based Composites Filled with Biochar Derived from Waste Coffee Grounds. *Polymers*, vol. *11*, p. 1-15.

Salmeia, Khalifah A., Gaan, Sabyasachi. & Malucelli, Giulio. (2016). Recent Advances for Flame Retardancy of Textiles Based on Phosphorus Chemistry. In: *Polymers*, vol. *8*, n. 319, pp. 1-36. - ISSN 2073-4360.

In: Low-Density Polyethylene  
Editor: Johan Geisler  
ISBN: 978-1-53618-192-0  
© 2020 Nova Science Publishers, Inc.

*Chapter 2*

# A Comprehensive Study on Biodegradation Process of Low Density Polyethylene by Microorganisms

*Seyyed Mojtaba Mousavi[1,\*], Maryam Zarei[2], Seyyed Alireza Hashemi[3], Wei-Hung Chiang[1,†], Ahmad Gholami[4] and Yasin Sadeghipoor[2]*

[1]Department of Chemical Engineering,
National Taiwan University of Science and Technology, Taiwan
[2]Pharmaceutical Sciences Research Center,
Shiraz University of Medical Sciences, Shiraz, Iran
[3]Department of Mechanical Engineering,
Center for Nanofibers and Nanotechnology,
National University of Singapore, Singapore
[4]Biotechnology Research Center and Department of Pharmaceutical Biotechnology, school of Pharmacy,
Shiraz University of Medical Science, Shiraz, Iran

---

[\*] Corresponding Author's E-mail: mousavi.nano@gmail.com.
[†] Corresponding Author's E-mail: whchiang@mail.ntust.edu.tw.

## ABSTRACT

In today's life polyethylene can be introduced as a resistant polymer to degradation. Since it has several biological and chemical properties it plays a critical role in the preparing process of different products like plastic bags. Polyethylene types can be classified into several groups such as Cross-Linked Polyethylene (XLPE), Linear Low-Density Polyethylene (LLDPE) High-Density Polyethylene (HDPE) and Low-Density Polyethylene (LDPE). Their density, number of branches and surface functional groups are different which can provide numerous applications. LDPE can be achieved by a high-pressure approach which creates short and long chains. Also, it can be stated that LDPE possesses a translucent structure with 50 to 65% crystallinity. From an economic point of view, LDPE is a polymer with acceptable flexibility, enough strength and satisfactory cost which can widely be applied in drug and food packaging. Regarding the environmental issues associated with LDPE waste, many efforts have accomplished in order to identify different microorganisms for degradation process. The present chapter will cover different topics, first the comprehensive research progress of LDPE is reviewed, and second the mechanisms of LDPE biodegradation are summarized. After that, different aspects like bio and oxo-degradable LDPE are presented. In two last parts, the effect of microorganisms on LDPE and products of this degradation with their level of toxicity are discussed respectively. Finally, to boost the improvement for LDPE, the major opportunities and future challenges are also explained.

**Keyword:** Low Density Polyethylene (LDPE), biodegradation process, bacteria, fungi

## 1. INTRODUCTION

During the last decades, the wide usage of plastics could alter the nature and definition of waste [1]. In today's life, the role of plastics in human life has been highlighted due to its different applications. One of the most important issues related to plastics is their high durability which can be very effective in terms of environmental conditions [2, 3]. Indeed plastics are introduced as long-chain polymers that produce on a significant level [4]. Recent studies on Plastic debris have proven that plastic generation is a very serious issue in the ecosystem. By increasing the deposition rate of plastics, a remarkable amount of this material is entered into the aqua system which is very harmful to both humans and animals [5-7]. Polyethylene can be introduced as a resistant polymer to degradation. Since it has several biological and chemical properties it plays a critical role in the preparing process of different products like plastic bags. Many publications have confirmed that polyethylene can block the bowels of marine mammals, birds, and fishes. Moreover, by ingesting or entangling polyethylene, many different types of fauna are exposed to risk [8-10]. Although different reports have proposed for biodegradation of polyethylene, the speed of this process under natural conditions is sorely slow [11-16]. So far many methods have been applied for removing LDPE from the environment; however, most of these methods are unacceptable and undesirable in terms of costs. Hence the researchers have confirmed that by using microorganisms better biodegradation can be achieved [17] (Table 1).

It is reported that some factors like great hydrophobicity of LDPE and its high molecular weight are the main obstacles of the biodegradation process [18]. Several polymer pretreatments like thermo-oxidation, chemical oxidation, and photo-oxidations can create additional C=O bonds which lead to accelerating the degradation of LDPE. These additional bonds can prepare polymer surfaces for further microorganisms attack. Seneviratne et al. have claimed that microorganisms create hydrophobic proteins and these proteins can simply attach to the surface of polymers. Also, the growth rate of fungi in comparison with bacteria is high which causes higher nutrient availability [19].

Although different properties of polyethylene have been assessed during the last years, LDPE biodegradation has gained lower attention. It is reported that the biodegradation process under normal conditions can occur at very low rate. Previous reports have proposed the biotic or abiotic classes of LDPE degradation. In biotic class, the performance of microorganisms is an essential factor to modify and change the polymer chain [39, 40]. Also, in the abiotic class, the deterioration can be caused by UV irradiation or high temperature.

## Table 1. Different microorganisms for polyethylene biodegradation

| Microorganisms | Strains | Reference number |
|---|---|---|
| Fungi | Verticillium Lecanii | [20] |
| Bacteria | Paenibacillus Macerans | [21] |
| Bacteria | Bacillus Amyloliquefaciens | [21] |
| Bacteria | Staphylococcus Xylosus | [21] |
| Bacteria | Arthrobacter Viscosus | [21] |
| Bacteria | Acinetobacter Baumannii | [21] |
| Bacteria | Micrococcus Luteus | [21] |
| Bacteria | Pseudomonas Fluorescens | [21] |
| Bacteria | Micrococcus Lylae | [21] |
| Bacteria | Staphylococcus Cohnii | [21] |
| Bacteria | Bacillus Thuringiensis | [21] |
| Fungi | Aspergillus Niger | [22] |
| Fungi | Phanerochaete Chrysosporium | [23] |
| Fungi | Aspergillus Versicolor | [20] |
| Bacteria | Bacillus Brevies | [24] |
| Bacteria | Arthrobacter spp | [25] |
| Bacteria | Pseudomonas spp | [25] |
| Fungi | Cladosporium Cladosporioides | [26] |
| Fungi | Mortierella Alpina | [27] |
| Fungi | Penicillum Pinophilum | [28] |
| Bacteria | Bacillus Halodenitrificans | [29] |
| Bacteria | Bacillus Cereus | [21, 29] |
| Bacteria | Bacillus Pumilus | [21, 29, 30] |
| Fungi | Mucor Circinelloides | [31] |
| Fungi | Chaetomium spp | [32] |
| Bacteria | Delftia Acidovorans | [33] |
| Fungi | Penicillum Simplicissimum | [34] |
| Fungi | Aspergillus Flavus | [27, 31] |
| Bacteria | Streptomyces Badius | [35] |
| Bacteria | Rhodococcus Ruber | [36] |
| Bacteria | Streptomyces Setonii | [35] |
| Fungi | Fusarium Redolens | [37] |
| Bacteria | Staphylococcus Epidermidis | [38] |
| Bacteria | Streptomyces Setonii | [35] |

Hakkarainen et al. have stated that only one of two aforementioned classes can be applied to degradation of LDPE, while in nature, both of them have been incorporated and therefore better responses are gained [41]. The detailed information of the abiotic class and its mechanism have been collected by different researchers [11, 27, 42, 43]. Since polyethylene is insoluble in aqueous systems and possesses few functional groups, the biotic process is faced with several limitations. Also, the high molecular weight of polyethylene can be an effective factor in the biodegradation process. In this chapter, an effort has been devoted to pool all the existing literature on biodegradation process of low density polyethylene under the following objectives: first to enlist all Mechanisms of LDPE biodegradation; second to highlight different aspects of bio and oxo-degradable LDPE; third to investigate the effect of microorganisms on degradation process and finally to explain conclusion and perspectives.

## 2. MECHANISMS OF LDPE BIODEGRADATION

Biodegradation process has been applied as incorporation of different items like LDPE properties, variation of organisms and intrinsic pretreatment. Several characteristics of polymers for instance molecular weight and presence of functional groups or additives are important factors in their degradation [44]. Since LDPE possesses very large molecules, it cannot cross into the membrane of cellular, hence the depolymerization is required. The first step in LDPE degradation is the conversion of this polymer to monomers and after that, mineralization occurs. Using this action, better absorption and higher biodegradation by the microbial cell can be achieved. It is reported that the presence of microorganisms like fungi can create large protuberance and also lead small-scale bursting [45]. Also, microbial enzymes lead to the deterioration of LDPE molecules by the two-step method. In this method, the enzymes attach to the LDPE and after that, they catalyze a hydrolytic segmentation. Intracellular and extracellular depolymerization processes have been applied by fungi in LDPE degradation. The hydrolysis of an endogenic carbon container by microbes has been referred to intracellular process. On the other hand in the extracellular process, the aforementioned mechanism is not surely occurred by microorganisms [46].

In the extracellular process, the enzymes can divide massive polymers into small chains like monomers. The produced monomers or dimers are able to cross the internal membranes (called depolymerization). In the next step, short chains are converted into other products like $CH_4$ and $CO_2$ (called mineralization). Many reports have demonstrated that these products can be applied as potential energy sources [11]. Also, it is verified that using UV light leads to a satisfactory decrease in molecular weight and acceptable formation of carbonyl groups [20, 47-49]. Since the estimation of LDPE biodegradation under different conditions is a valuable issue, several approaches are available for this purpose [50]. In terms of biochemical process, oxidation and reduction of molecular weight are two important factors in LDPE biodegradation. Yoon et al. have confirmed that the reduction of molecular weight is an essential step for transporting molecules by the cell membranes [51]. They also claimed that different microorganisms usually attack material with specific molecular weights.

As demonstrated in Figure 1 after reduction of molecular weight, the oxidation process is needed for converting the hydrocarbon into carboxylic acid [52]. Synergistic interactions between abiotic and biotic groups can lead to a reduction of molecular weight and oxidation processes. The group of enzymes that can diminish the LDPE molecules is called the biotic group and it has received less attention during the last years. In 2016 a group of researchers has proposed that the incubation of certain enzymes can affect the molecular weight and also ketocarbonyl index of LDPE [53]. Rojo et al. have demonstrated that alkane hydroxylase provides the oxidation and consequently better degradation of hydrocarbon occurs [54].

Biodegradable plastics have been introduced to surmount waste problems. LDPE is one of the main biodegradable materials which follows the criteria of industrial applications. The direct degradation process of LDPE has been proposed by previous researchers [55-57]. They claimed that the role of bacterial and fungal strains cannot be ignored in LDPE biodegradation. It is reported that microorganisms can preserve nature under low pH conditions and low nutrient availability [58]. Oxo-degradable plastics possess certain additives such as manganese, cobalt or iron that can expedite the degradation process. The researchers claimed that the rise in carbonyl groups and the decrease of molecular weight accelerate biodegradation process. Oxidation with nitric acid, using UV light and exposure to heat can be considered as main initiators for the segmentation of LDPE. Oxo-degradable LDPE provides new insights to overcome present environmental issues. The properties of bio and oxo-degradable LDPEs are not the same, for instance by exposure to oxygen the larger molecules break down to smaller ones in oxo-degradable LDPE. Even these small molecules can cause serious problems by distributing on the earth or being consumed by both humans and animals. Different oxo-biodegradable additives can be seen in many products like polyethylene terephthalate, polypropylene, and polyvinyl chloride [59].

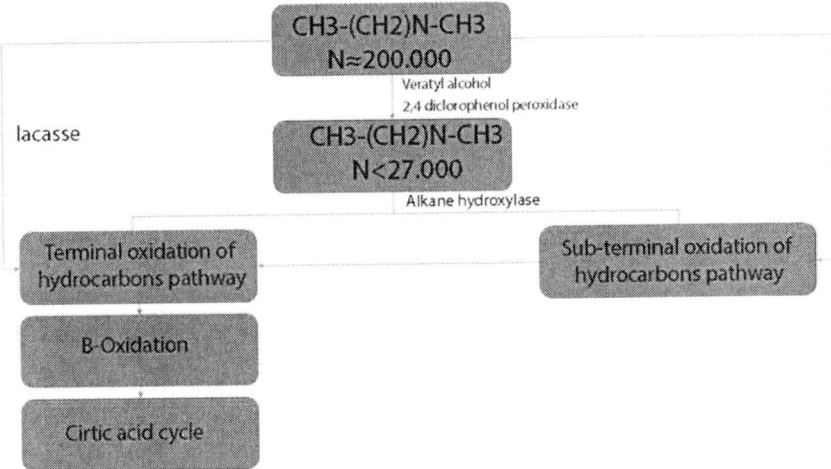

Figure 1. Main mechanisms of LDPE degradation [35, 52, 54].

## 3. EFFECT OF MICROORGANISMS (BACTERIA AND FUNGI) ACTIVITY ON LDPE

A major component of waste is LDPE which has been broadly applied in different forms and applications. Thence, considerable researches have been devoted to finding an appropriate way to degrade this material. Since the biodegradability of LDPE doesn't have a typical strategy, several methods have been proposed to elucidate this action. For instance, many researchers have claimed that applying pure strains can impressively lead to degrading LDPE [35, 60, 61]. According to the results, the researchers concluded that by using pure strains better estimation and high-quality evaluation of LDPE degradation can be achieved. In addition, by this strategy, an acceptable separation between biological and chemical degradation of LDPE has been provided. Therefore the impact of various bacteria like *Bacillus amyloliquefaciens* [62], *Rhodococcus ruber* [36], and *Acinetobacter buammi* [63] on LDPE biodegradation had been assessed. One disadvantage of the mentioned strategy was that some probabilities about the interaction between strains were ignored which lead to decrease accuracy. Therefore, the combination of microbial communities has been introduced as a potential alternative for the previous strategy. A group of researchers has demonstrated that a mixture of *Brevibacillus borstelensis* and *Bacillus species* can be regarded as an effective strategy in LDPE biodegradation (Figure 2). They first prepared a scaffold with variable temperature, to verifying the impact of temperature and time on biodegradation of LDPE. The results of their experiment have confirmed that using the mixture does not always lead to further biodegradation.

As stated before, the production of bacterial mixture for eco-friendly and beneficial degradation has gained considerable attention during the last years. It is necessary to mention that this strategy can improve the detoxification rate and also enhance the degradation process (Table 2). Skariyachan et al. tried to formulate a new microbial mixture that could enhance the degradation of LDPE. They designed the bacterial mixture of *Pseudomonas putida*, *Bacillus subtilis*, and *Pseudomonas stutzeri*, and finally, they have concluded that a significant fraction of LDPE was degraded by this novel consortia [65]. Das et al. have demonstrated that two bacterial isolates *Bacillus* strains can lead to LDPE biodegradation. They measured the pH value, weight reduction, and $CO_2$ content to proving their observations (Figure 3) [62].

## Table 2. Bacteria and biodegradation result in LDPE biodegradation

| Source | Bacteria | Exposure time | Obtained results | Refs |
|---|---|---|---|---|
| Water | *Bacillus sphericus* | 12 months | Weight loss from 9% to 19% | [77] |
| Dumped soil area | *Arthobacter defluvii* | 1 month | Weight loss from 20% to 30% | [78] |
| ATCC | *Rhodococcus rhorocuros* | 6 months | 50% mineralization | [26] |
| Water | *Streptomyces* spp. | 1 month | Slight weight loss | [70] |
| Agricultural waste | *Rhodococcus ruber* | 2 months | 0.86% weight loss per week | [79] |
| Soil | *Delftia* sp. | 3 months | Changing in chemical properties | [80] |
| Gene bank | *Pseudomonas* sp. | 2 months | 50.5% weight loss | [81] |
| Soil | *Trichoderma* spp. | 1 month | 7.5% weight loss | [82] |
| Waste water | *Pseudomonas* sp. | 1 month | 40% weight loss | [83] |
| Plastic debris in soil | *Rhodococcus* sp. | 1 months | Slight weight loss | [84] |
| Soil | *Rhodococcus ruber* | 1 month | Up to 8% weight loss | [85] |
| ATCC 29672 | *Rhodococcus rhorocuros* | 6 months | Various mineralization | [47] |
| Waste coal | *Bacillus* ssp. | 7 months | Changing in mechanical properties | [21] |
| Soil | *Bacillus cereus* | 3 months | 7%–10% mineralization | [64] |
| Soil | *Listeria* | 1 month | 10% weight loss | [86] |
| MCC No. 2183 | *Bacillus subtilis* | 1 month | 9.26% weight loss | [87] |
| Water | *Bacillus subtillis* | 1 month | 1.5%–1.75% weight loss | [88] |
| DSMZ | *Brevibacillus borstelensis* | 3 months | 17% weight loss | [60] |
| Farmland soil | *Lysinibacillus* sp. | 4 months | 29.5% weight loss | [89] |
| Municipal landfill | *Acinetobacter bumannii* | 1 month | Biomass production | [63] |
| Soil waste | *Bacillus amyloliquefaciens* | 2 months | 15% mineralization | [90] |
| Beach soil | *Pseudomonas* sp. | 3 months | 4.9%–28.6% $CO_2$ production | [91] |
| ATCC | *Pseudomonas putida* | 4 months | 9%–20% weight loss | [92] |
| Mineral oil | *Pseudomonas* sp. | 2 month | 5% weight loss | [61] |

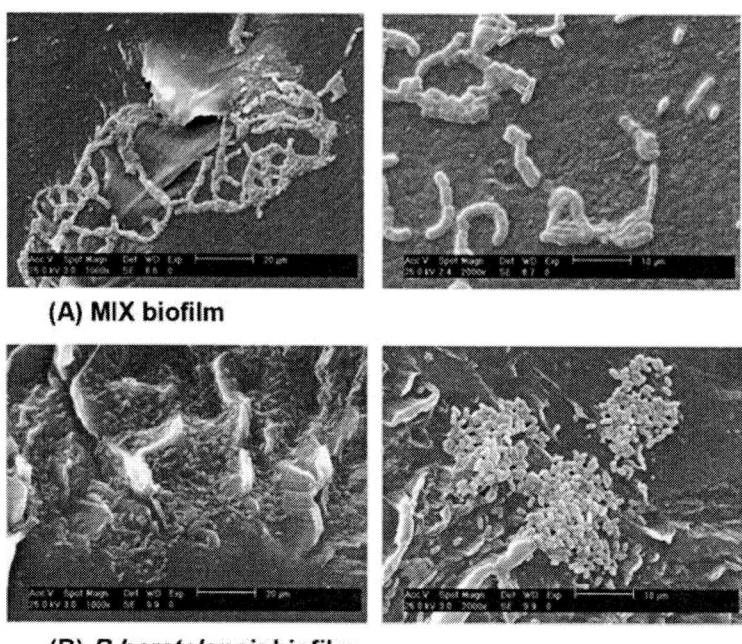

(A) MIX biofilm

(B) *B.borstelensis* biofilm

Figure 2. ESEM images of (A) MIX biofilm (B) *B. borstelensis* biofilm [64].

Both *Bacillus amyloliquefaciens* (BSM-1) and *Bacillus amyloliquefaciens* (BSM-2) produced different enzymes, which pH variation could confirm this metabolic activity. Based on the results they stated that BSM-2 had a better degradation pathway in comparison with BSM-1 which could degrade LDPE in a short period of time. The polymers [66, 67] have been introduced as major carbon sources for growing the organisms on the surface. The degradation process starts with cleaving larger molecules into dimers, oligomers or monomers [68]. It is reported that some extracellular enzymes like glucosidases, serine hydrolase, cutinasem, catalase, and manganese peroxidase have been secreted by *Aspergillus flavus*, *Pestalotiopsis microspore*, *Aspergillus oryzae*, *Aspergillus niger*, and *Phanerochaete chrysosporium* respectively. Indeed these produced compounds can be utilized as potential energy and carbon source. Different health and environmental consequences will occur, if the residual small molecules are not completely consumed by the microorganisms [68]. Since fungi can produce essential degrading enzymes [69] and even extracellular macromolecules, they are widely applied in LDPE biodegradation. Some extracellular polymers like polysaccharides can cause colonization of the polymer surface. Previous investigations have confirmed that using fungal strains like *Phanerochaete chrysosporium* [50], *Aspergillus. flavus* [70], *A. niger* [28, 71] and *A. fumigatus* [72] is an effective strategy for LDPE biodegradation. Jayanthi et al. have compared the degradation of LDPE by *Streptomyces* sp and *Aspergillus nomius* in the absence and presence of glycerol. They stated that in degradation process, LDPE has been consumed by *Streptomyces* sp and *Aspergillus nomius* which leads degradation with high efficiency (Figure 4). Fungal strains have been survived under low nutrient and harsh conditions and they possess valuable oxidative enzymes. Also, fungi can simply penetrate crevices and holes, thus they are promising candidates in LDPE degradation. A group of researchers have evaluated the biodegradation of LDPE by a mixture of *Aspergillus sp.* and *pseudomonas aeruginosa* under mathematical models. Based on their result, it is obvious that an effective β-oxidation of LDPE can be achieved by fungal strains [73]. According to the previous literature, it can be concluded that the degradation of LDPE by fungi has several limitations. One of these limitations is that fungi can grow only on the external surface of LDPE and the performance of fungi will decrease when it penetrates into the deep depth. While the penetration of bacteria even in the backbone structure is acceptable. Another limitation is that the presence of fungi on the surface can insulate the structural cells from oxygen and nutrient materials. This insulating impact may lead to some variations in LDPE biodegradations [70]. Kim et al. have stated that the faster penetration of fungi in comparison with bacteria and also the growth extension of hyphae are two important reasons for selecting fungal versus bacterial strains in LDPE biodegradation [74]. Nayak et al. have used microorganisms in the decomposition of LDPE in both *in vivo* and *in vitro* conditions. They stated that fungal strains can form hydrophobic proteins and effective enzymes [75]. A group of researchers has demonstrated that without any photothermal treatments and only by using two fungal strains (*A. niger* and *A. terreus),* biodegradation of LDPE occurs. They tried to decompose LDPE using sucrose as a source under the least nutritional conditions. The significant weight loss (about 30%) has been obtained by fungi that can confirm the effective performance of fungi species even in low nutritional conditions [76]. Figure 5 has illustrated the SEM micrographs of *Aspergillus niger* and *Aspergillus terreus* over LDPE degradation.

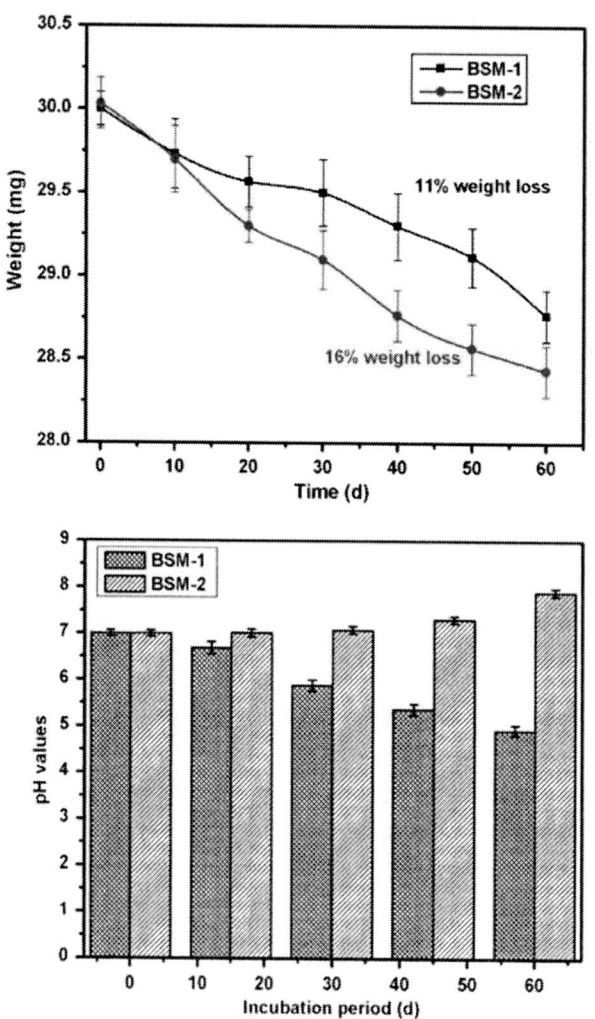

Figure 3. (Top) Weight loss of LDPE, (bottom) pH variation during degradation by microbes [62].

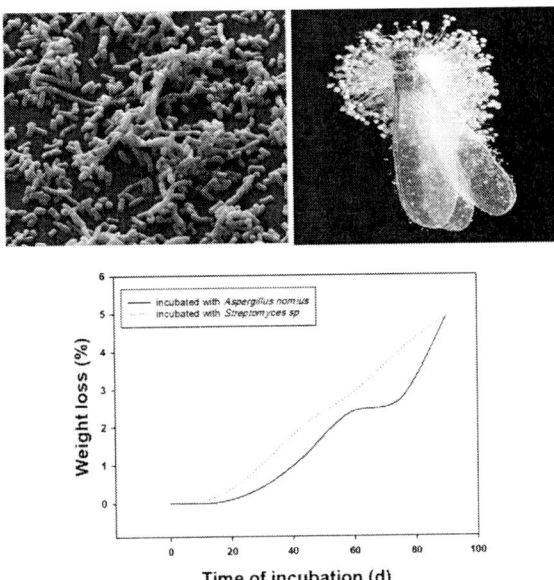

Figure 4. Image of Streptomyces *sp* (Left), *Aspergillus nomius* (Right) and graph of weight loss in LDPE biodegradation.

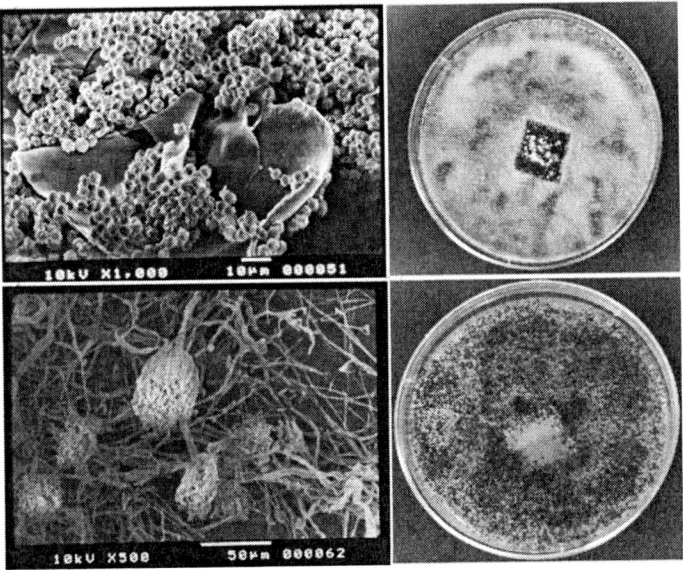

Figure 5. SEM images and fungi growth of (top) *Aspergillus niger* and (bottom) *Aspergillus terreus* over LDPE degradation [76].

## 4. PRODUCTS AND MECHANICAL PROPERTIES OF LDPE DEGRADATION

So far many efforts have been devoted to optimizing the biodegradation of LDPE [93]. During the decomposition process, the emission of $CO_2$ gas cannot be ignored [19, 47]. Sivan and his coworkers reported that LDPE acts as a carbon source in biodegradation by Rhodococcus rubber and also proteins and polysaccharides are by-products of this process [94]. It is necessary to mention that other products like Betamethasone acetate and Ergosta-5, 1-Monanalinoeoglycerol trimethylsilyl ether have been produced in the decomposition of LDPE. Kyaw et al. have identified several products of LDPE biodegradation, for example, Hexacosane, methyl, Docosane, Tricosane, 2-Benxenedicarboxylic acid, Tetrachloroethylene, Octadecane, 3-Chloropropionic acid, Eicosane, and Butyl ester. From the environmental point of view, the toxicity of LDPE biodegraded products is a critical issue. Hence the researchers have tried to determine the toxicity of these products, for instance, Aswale [95] in her experiment, tested the toxicity of products both on plant and animal systems. She applied these products on different seeds like *Carthamus tinctorius, Glycine max, Arachis hypogaea* and *Helianthus annuus*. Based on the results, an almost low decrease in the germination of the aforementioned seeds was seen. Also, it is reported that LDPE biodegraded products could affect the mortality rates of *Chironomous larvae*.

Table 3. Variation in mechanical properties of LDPE caused by microbial effects

| Conditions | Exposure time | %Δ Tensile strength | %Δ Elongation | Refs |
|---|---|---|---|---|
| Beach soil | 7 months | -19.5 | -1.5 | [21] |
| Sea water | 7 months | -16.4 | 4 | [21] |
| Mineral media | 1 month | -30 | NR | [77] |
| Forest soil | 7 months | -16.4 | -4 | [21] |
| Sea water (sterile) | 12 months | -3.8 | 2.7 | [77] |
| Sea water | 12 months | -15 | -12 | [96] |

According to previous literature, it is accepted that in LDPE biodegradation, the breaking load may occur. Also, the researchers have demonstrated that some oxidation changes in structure and crystallinity leads to modifying mechanical properties. In Table 3 some of the mechanical properties of LDPE after biodegradation have been described. LDPE biodegradation under soil conditions causes remarkable changes in its mechanical properties (i.e., elongation at break of 98%). It is reported that in soil with rich organic matter, a faster biodegradation process can be occurred. Nowak et al. claimed that some fungal strains like *Aspergillus awamori, Gliocladium viride*, and *Mortierella subtilissima* cause colonization in LDPE [21]. The researchers utilize a universal mechanical testing system to determine the changes in mechanical properties while in the past rheological analysis was being used for the determination of loss modulus of the polymer. However, it seems that the effects of microorganisms such as fungi and bacteria can have relatively low impact on the mechanical properties of LDPE [96-99].

## CONCLUSION AND PERSPECTIVES

Current researches performed in the field of LDPE biodegradation, using pure and mixed strains have verified that the biodegradation process of LDPE is not a fast mechanism in nature. Indeed, LDPE biodegradation depends on some abiotic factors like oxidizing agents. Also, the presence of UV light, temperature variations and physicochemical properties can affect the biodegradation rate. Many factors like the degree of oxidation, structure, and crystallinity of polymers have been considered as major factors in the decomposition process. Hence the biodegradation of LDPE is defined by several steps like biofragmentation, mineralization, biodeterioration, and bioassimilation. Although many efforts have been devoted to introducing mineralization and biofragmentation, more evidence of biodeterioration and bioassimilation are required. Degradation of LDPE by microorganisms is one of the most essential issues in which there are still vague points. For example, it has been proven that LDPE morphology can specify the amount of degradation. Although amorphous regions simply decompose the LDPE, no reports have been found based on strong crystalline areas that are susceptible to fungal and bacterial attack. Also, the exposure time and the decomposition rate of LDPE degradation are of high importance which have been underestimated in previous literature. This chapter has covered the background and major environmental concerns about LDPE. Also, the different aspects of bio and oxo-degradable LDPE, the effect of microorganisms and products of the biodegradation process have been reported. Although mechanisms of LDPE biodegradation have been reviewed here, future studies should be followed to optimize this process.

## REFERENCES

[1] Sheavly, S. B., *Sixth Meeting of the UN Open-ended Informal Consultative Process on Oceans & the Law of the Sea.* 2005.

[2] Brown, B. L., et al., Functional and anatomical outcomes of facial nerve injury with application of polyethylene glycol in a rat model. *JAMA facial plastic surgery,* 2019. 21(1): p. 61-68.

[3] Muhammad, U. L., I. M. Shamsuddin, and K. M. Yahaya, Environmental Impacts of Waste Disposal: An Overview on the Disposal of Polyethylene Bags in Gusau City Zamfara State. *Journal of Evolutionary Science,* 2019. 1(2): p. 31.

[4] Scott, G., *Polymers and the Environment.* 1999: Royal Society of Chemistry.

[5] Frias, J., P. Sobral, and A. Ferreira, Organic pollutants in microplastics from two beaches of the Portuguese coast. *Marine Pollution Bulletin,* 2010. 60(11): p. 1988-1992.

[6] Simon-Walker, R., et al., Hemocompatibility of hyaluronan enhanced linear low density polyethylene for blood contacting applications. *Journal of Biomedical Materials Research Part B: Applied Biomaterials,* 2018. 106(5): p. 1964-1975.

[7]  Teuten, E. L., et al., Transport and release of chemicals from plastics to the environment and to wildlife. *Philosophical Transactions of the Royal Society B: Biological Sciences*, 2009. 364(1526): p. 2027-2045.

[8]  MacDonald, D. W., et al., Oxidation, damage mechanisms, and reasons for revision of sequentially annealed highly crosslinked polyethylene in total knee arthroplasty. *The Journal of arthroplasty*, 2018. 33(4): p. 1235-1241.

[9]  Secchi, E. R. and S. Zarzur, *Plastic debris ingested by a Blainville's beaked whale, Mesoplodon densirostris, washed ashore in Brazil.* 1999.

[10] Spear, L. B., D. G. Ainley, and C. A. Ribic, Incidence of plastic in seabirds from the tropical pacific, 1984–1991: relation with distribution of species, sex, age, season, year and body weight. *Marine Environmental Research*, 1995. 40(2): p. 123-146.

[11] Gu, J. D., Microbiological deterioration and degradation of synthetic polymeric materials: recent research advances. *International biodeterioration & biodegradation*, 2003. 52(2): p. 69-91.

[12] Hakkarainen, M. and A. C. Albertsson, Environmental degradation of polyethylene, in *Long term properties of polyolefins*. 2004, Springer. p. 177-200.

[13] Koutný, M., J. Lemaire, and A. M. Delort, Biodegradace polyethylenových filmů's prooxidanty. *Chemosphere*, 2006.

[14] Arutchelvi, J., et al., *Biodegradation of polyethylene and polypropylene.* 2008.

[15] Mukherjee, S., U. Roy Chaudhuri, and P. P. Kundu, Biodegradation of polyethylene via complete solubilization by the action of *Pseudomonas fluorescens*, biosurfactant produced by *Bacillus licheniformis* and anionic surfactant. *Journal of Chemical Technology & Biotechnology*, 2018. 93(5): p. 1300-1311.

[16] Xu, J., et al., Polyaniline modified mesoporous titanium dioxide that enhances oxo-biodegradation of polyethylene films for agricultural plastic mulch application. *Polymer International*, 2019. 68(7): p. 1332-1340.

[17] Gholami, A., et al., Industrial production of polyhydroxyalkanoates by bacteria: opportunities and challenges. *Minerva Biotecnologica*, 2016. 28(1): p. 59-74.

[18] Hamid, S. H., *Handbook of polymer degradation*. 2000: CRC Press.

[19] Seneviratne, G., et al., *Polyethylene biodegradation by a developed Penicillium-Bacillus biofilm*. 2006.

[20] Karlsson, S., O. Ljungquist, and A. C. Albertsson, Biodegradation of polyethylene and the influence of surfactants. *Polymer degradation and stability*, 1988. 21(3): p. 237-250.

[21] Nowak, B., et al., Microorganisms participating in the biodegradation of modified polyethylene films in different soils under laboratory conditions. *International biodeterioration & biodegradation*, 2011. 65(6): p. 757-767.

[22] Raghavan, D. and A. Torma, DSC and FTIR characterization of biodegradation of polyethylene. *Polymer Engineering & Science*, 1992. 32(6): p. 438-442.

[23] Manzur, A., F. Cuamatzi, and E. Favela, Effect of the growth of *Phanerochaete chrysosporium* in a blend of low density polyethylene and sugar cane bagasse. *Journal of applied polymer science*, 1997. 66(1): p. 105-111.

[24] Watanabe, T., et al., Biodegradability and degrading microbes of low-density polyethylene. *Journal of applied polymer science*, 2009. 111(1): p. 551-559.

[25] Balasubramanian, V., et al., High-density polyethylene (HDPE)-degrading potential bacteria from marine ecosystem of Gulf of Mannar, India. *Letters in applied microbiology*, 2010. 51(2): p. 205-211.

[26] Bonhomme, S., et al., Environmental biodegradation of polyethylene. *Polymer Degradation and Stability*, 2003. 81(3): p. 441-452.

[27] Koutny, M., et al., Acquired biodegradability of polyethylenes containing pro-oxidant additives. *Polymer degradation and stability*, 2006. 91(7): p. 1495-1503.

[28] Volke-Sepúlveda, T., et al., Thermally treated low density polyethylene biodegradation by *Penicillium pinophilum* and

*Aspergillus niger. Journal of applied polymer science*, 2002. 83(2): p. 305-314.

[29] Roy, P., et al., Degradation of abiotically aged LDPE films containing pro-oxidant by bacterial consortium. *Polymer degradation and stability*, 2008. 93(10): p. 1917-1922.

[30] Satlewal, A., et al., Comparative biodegradation of HDPE and LDPE using an indigenously developed microbial consortium. *J Microbiol Biotechnol*, 2008. 18(3): p. 477-482.

[31] Pramila, R. and K. V. Ramesh, Biodegradation of low density polyethylene (LDPE) by fungi isolated from marine water a SEM analysis. *Afr J Microbiol Res*, 2011. 5(28): p. 5013-5018.

[32] Sowmya, H., M. Ramalingappa, and M. Krishnappa, Degradation of polyethylene by *Chaetomium* sp. and *Aspergillus flavus*. *Int. J. Recent Sci. Res*, 2012. 3(513): p. e517.

[33] Koutny, M., et al., Soil bacterial strains able to grow on the surface of oxidized polyethylene film containing prooxidant additives. *International Biodeterioration & Biodegradation*, 2009. 63(3): p. 354-357.

[34] Yamada-Onodera, K., et al., Degradation of polyethylene by a fungus, *Penicillium simplicissimum* YK. *Polymer degradation and stability*, 2001. 72(2): p. 323-327.

[35] Pometto, A., B. Lee, and K. E. Johnson, Production of an extracellular polyethylene-degrading enzyme (s) by Streptomyces species. *Appl. Environ. Microbiol.*, 1992. 58(2): p. 731-733.

[36] Orr, I. G., Y. Hadar, and A. Sivan, Colonization, biofilm formation and biodegradation of polyethylene by a strain of *Rhodococcus ruber*. *Applied microbiology and biotechnology*, 2004. 65(1): p. 97-104.

[37] Albertsson, A. C., The shape of the biodegradation curve for low and high density polyethenes in prolonged series of experiments. *European Polymer Journal*, 1980. 16(7): p. 623-630.

[38] Chatterjee, S., et al., Enzyme-mediated biodegradation of heat treated commercial polyethylene by Staphylococcal species. *Polymer Degradation and Stability*, 2010. 95(2): p. 195-200.

[39] Mousavi, S. M., et al., Pb (II) removal from synthetic wastewater using Kombucha Scoby and graphene oxide/Fe3O4. *Physical Chemistry Research*, 2018. 6(4): p. 759-771.

[40] Mousavi, S. M., et al., Data on cytotoxic and antibacterial activity of synthesized Fe3O4 nanoparticles using *Malva sylvestris*. *Data in brief*, 2020. 28: p. 104929.

[41] Lindström, A., A. C. Albertsson, and M. Hakkarainen, Quantitative determination of degradation products an effective means to study early stages of degradation in linear and branched poly (butylene adipate) and poly (butylene succinate). *Polymer degradation and stability*, 2004. 83(3): p. 487-493.

[42] Mazerolles, T., et al., Development of co-continuous morphology in blends of thermoplastic starch and low-density polyethylene. *Carbohydrate polymers*, 2019. 206: p. 757-766.

[43] Muhammad, D., Z. Ahmad, and N. Aziz, Low density polyethylene tubular reactor control using state space model predictive control. *Chemical Engineering Communications*, 2019: p. 1-17.

[44] Gu, J. D., Biofouling and prevention: corrosion, biodeterioration and biodegradation of materials, in *Handbook of Environmental Degradation of Materials*. 2005, Elsevier. p. 179-206.

[45] Griffin, G., Synthetic polymers and the living environment. *Pure and Applied Chemistry*, 1980. 52(2): p. 399-407.

[46] Tokiwa, Y. and B. P. Calabia, Review degradation of microbial polyesters. *Biotechnology letters*, 2004. 26(15): p. 1181-1189.

[47] Fontanella, S., et al., Comparison of the biodegradability of various polyethylene films containing pro-oxidant additives. *Polymer Degradation and Stability*, 2010. 95(6): p. 1011-1021.

[48] Alshehrei, F., Biodegradation of low density polyethylene by fungi isolated from Red sea water. *International Journal of Current Microbiology and Applied Sciences*, 2017. 6: p. 1703-1709.

[49] Laha, S. D., K. Dutta, and P. P. Kundu, Biodegradation of Low Density Polyethylene Films, in *Handbook of Research on Microbial Tools for Environmental Waste Management*. 2018, IGI Global. p. 282-318.

[50] Orhan, Y., J. Hrenovic, and H. Buyukgungor, Biodegradation of plastic compost bags under controlled soil conditions. *Acta Chimica Slovenica*, 2004. 51(3): p. 579-588.

[51] Yoon, M. G., H. J. Jeon, and M. N. Kim, Biodegradation of polyethylene by a soil bacterium and AlkB cloned recombinant cell. *J Bioremed Biodegrad*, 2012. 3(4): p. 1-8.

[52] Albertsson, A. C., S. O. Andersson, and S. Karlsson, The mechanism of biodegradation of polyethylene. *Polymer degradation and stability*, 1987. 18(1): p. 73-87.

[53] Gray, N., et al., Influence of cellulose nanocrystal on strength and properties of low density polyethylene and thermoplastic starch composites. *Industrial Crops and Products*, 2018. 115: p. 298-305.

[54] Rojo, F., Enzymes for aerobic degradation of alkanes. *Handbook of hydrocarbon and lipid microbiology*, 2010. 2: p. 781-797.

[55] Gajendiran, A., S. Krishnamoorthy, and J. Abraham, Microbial degradation of low-density polyethylene (LDPE) by Aspergillus clavatus strain JASK1 isolated from landfill soil. 3 *Biotech*, 2016. 6(1): p. 52.

[56] Lukanina, Y. K., et al. Oxo-degradation of LDPE with pro-oxidant additive. In *IOP Conference Series: Materials Science and Engineering*. 2019. IOP Publishing.

[57] Montazer, Z., M. B. Habibi Najafi, and D. B. Levin, Microbial degradation of low-density polyethylene and synthesis of polyhydroxyalkanoate polymers. *Canadian journal of microbiology*, 2019. 65(3): p. 224-234.

[58] Delacuvellerie, A., et al., The plastisphere in marine ecosystem hosts potential specific microbial degraders including *Alcanivorax borkumensis* as a key player for the low-density polyethylene degradation. *Journal of hazardous materials*, 2019. 380: p. 120899.

[59] Zimmerman, J. B., et al., *Global stressors on water quality and quantity*. 2008, ACS Publications.

[60] Hadad, D., S. Geresh, and A. Sivan, Biodegradation of polyethylene by the thermophilic bacterium *Brevibacillus borstelensis*. *Journal of applied microbiology*, 2005. 98(5): p. 1093-1100.

[61] Tribedi, P. and A. K. Sil, Low-density polyethylene degradation by *Pseudomonas* sp. AKS2 biofilm. *Environmental Science and Pollution Research*, 2013. 20(6): p. 4146-4153.

[62] Das, M. P. and S. Kumar, An approach to low-density polyethylene biodegradation by *Bacillus amyloliquefaciens*. *3 Biotech*, 2015. 5(1): p. 81-86.

[63] Pramila, R. and K. V. Ramesh, Potential biodegradation of low density polyethylene (LDPE) by *Acinetobacter baumannii*. *African Journal of Bacteriology Research*, 2015. 3(1): p. 92-95.

[64] Abrusci, C., et al., Biodegradation of photo-degraded mulching films based on polyethylenes and stearates of calcium and iron as pro-oxidant additives. *International Biodeterioration & Biodegradation*, 2011. 65(3): p. 451-459.

[65] Skariyachan, S., et al., Novel bacterial consortia isolated from plastic garbage processing areas demonstrated enhanced degradation for low density polyethylene. *Environmental Science and Pollution Research*, 2016. 23(18): p. 18307-18319.

[66] Mousavi, S., et al., Nanosensors for Chemical and Biological and Medical Applications. *Med Chem* (Los Angeles), 2018. 8(8): p. 2161-0444.1000515.

[67] Mousavi, S. M., et al., A conceptual review of rhodanine: current applications of antiviral drugs, anticancer and antimicrobial activities. *Artificial cells, nanomedicine, and biotechnology*, 2019. 47(1): p. 1132-1148.

[68] Awasthi, S., et al., Biodegradation of thermally treated low density polyethylene by fungus *Rhizopus oryzae* NS 5. *3 Biotech*, 2017. 7(1): p. 73.

[69] Shah, A. A., et al., Biological degradation of plastics: a comprehensive review. *Biotechnology advances*, 2008. 26(3): p. 246-265.

[70] El-Shafei, H. A., et al., Biodegradation of disposable polyethylene by fungi and Streptomyces species. *Polymer degradation and stability*, 1998. 62(2): p. 361-365.

[71] Manzur, A., M. Limón-González, and E. Favela-Torres, Biodegradation of physicochemically treated LDPE by a consortium

of filamentous fungi. *Journal of Applied Polymer Science*, 2004. 92(1): p. 265-271.

[72] Gajendiran, A., S. Subramani, and J. Abraham. Effect of Aspergillus versicolor strain JASS1 on low density polyethylene degradation. In *IOP. Conf. Ser. Mater. Sci. Eng.* 2017.

[73] Kawai, F., et al., Comparative study on biodegradability of polyethylene wax by bacteria and fungi. *Polymer degradation and stability*, 2004. 86(1): p. 105-114.

[74] Kim, D. and Y. Rhee, Biodegradation of microbial and synthetic polyesters by fungi. *Applied microbiology and biotechnology*, 2003. 61(4): p. 300-308.

[75] Nayak, P. and A. Tiwari, Biodegradation of polythene and plastic by the help of microbial tools: a recent approach. *IJBAR*, 2011. 2(9): p. 344-355.

[76] Sáenz, M., et al., Minimal Conditions to Degrade Low Density Polyethylene by *Aspergillus terreus* and *niger*. *Journal of Ecological Engineering*, 2019. 20(6).

[77] Sudhakar, M., et al., Marine microbe-mediated biodegradation of low- and high-density polyethylenes. *International Biodeterioration & Biodegradation*, 2008. 61(3): p. 203-213.

[78] Thakur, P., *Screening of plastic degrading bacteria from dumped soil area*. 2012.

[79] Santo, M., R. Weitsman, and A. Sivan, The role of the copper-binding enzyme–laccase–in the biodegradation of polyethylene by the actinomycete *Rhodococcus ruber*. *International Biodeterioration & Biodegradation*, 2013. 84: p. 204-210.

[80] Peixoto, J., L. P. Silva, and R. H. Krüger, Brazilian Cerrado soil reveals an untapped microbial potential for unpretreated polyethylene biodegradation. *Journal of hazardous materials*, 2017. 324: p. 634-644.

[81] Rajandas, H., et al., A novel FTIR-ATR spectroscopy based technique for the estimation of low-density polyethylene biodegradation. *Polymer Testing*, 2012. 31(8): p. 1094-1099.

[82] Hikmah, M., R. Setyaningsih, and A. Pangastuti. The Potential of Lignolytic Trichoderma Isolates in LDPE (Low Density Polyethylene) Plastic Biodegradation. In *IOP Conference Series*: *Materials Science and Engineering*. 2018. IOP Publishing.

[83] Nanda, S. and S. S. Sahu, Biodegradability of polyethylene by *Brevibacillus*, *Pseudomonas*, and *Rhodococcus* spp. *New York Science Journal*, 2010. 3(7): p. 95-98.

[84] Rodríguez-Seijo, A., et al., Low-density polyethylene microplastics as a source and carriers of agrochemicals to soil and earthworms. *Environmental Chemistry*, 2019. 16(1): p. 8-17.

[85] George, R. M., *Isolation and characterization of low-density polyethylene degrading and biosurfactant-producing bacteria from soils.* 2019, Brac University.

[86] Raaman, N., et al., Biodegradation of plastic by Aspergillus spp. isolated from polythene polluted sites around Chennai. *J Acad Indus Res*, 2012. 1(6): p. 313-316.

[87] Vimala, P. and L. Mathew, Biodegradation of Polyethylene using Bacillus subtilis. *Procedia Technology*, 2016. 24: p. 232-239.

[88] Cardoso, C. F., J. d. A. F. Faria, and E. H. M. Walter, Modeling of sporicidal effect of hydrogen peroxide in the sterilization of low density polyethylene film inoculated with Bacillus subtilis spores. *Food control*, 2011. 22(10): p. 1559-1564.

[89] Esmaeili, A., et al., Biodegradation of low-density polyethylene (LDPE) by mixed culture of Lysinibacillus xylanilyticus and *Aspergillus niger* in soil. *Plos one*, 2013. 8(9).

[90] Das, M. P. and S. Kumar, Influence of cell surface hydrophobicity in colonization and biofilm formation on LDPE biodegradation. *Int J Pharm Pharm Sci*, 2013. 5(4): p. 690-4.

[91] Pathak, V. M. and N. Kumar, Implications of $SiO_2$ nanoparticles for in vitro biodegradation of low-density polyethylene with potential isolates of *Bacillus*, *Pseudomonas*, and their synergistic effect on *Vigna mungo* growth. *Energy, Ecology and Environment*, 2017. 2(6): p. 418-427.

[92] Kyaw, B. M., et al., Biodegradation of low density polythene (LDPE) by Pseudomonas species. *Indian journal of microbiology*, 2012. 52(3): p. 411-419.

[93] Mousavi, S., et al., Improved morphology and properties of nanocomposites, linear low density polyethylene, ethylene-co-vinyl acetate and nano clay particles by electron beam. *Polymers from Renewable Resources*, 2016. 7(4): p. 135-153.

[94] Sivan, A., M. Szanto, and V. Pavlov, Biofilm development of the polyethylene-degrading bacterium *Rhodococcus ruber*. *Applied microbiology and biotechnology*, 2006. 72(2): p. 346-352.

[95] Aswale, P. N., *Studies on biodegradation of polythene*. 2011.

[96] Pegram, J. E. and A. L. Andrady, Outdoor weathering of selected polymeric materials under marine exposure conditions. *Polymer Degradation and Stability*, 1989. 26(4): p. 333-345.

[97] Mousavi, S. M., et al., Modification of phenol novolac epoxy resin and unsaturated polyester using sasobit and silica nanoparticles. *Polymers from Renewable Resources*, 2017. 8(3): p. 117-132.

[98] Vickers, N. J., Animal Communication: When I'm Calling You, Will You Answer Too? *Current Biology*, 2017. 27(14): p. R713-R715.

[99] Mousavi, S., M. Zarei, and S. Hashemi, Polydopamine for Biomedical Application and Drug Delivery System. *Med. Chem* (Los Angeles), 2018. 8: p. 218-229.

In: Low-Density Polyethylene
Editor: Johan Geisler

ISBN: 978-1-53618-192-0
© 2020 Nova Science Publishers, Inc.

Chapter 3

# BIOTRANSFORMATION RENDERED BY MICROBES ON VARIOUS LDPE

## *Anushree Suresh and Jayanthi Abraham*[*]

*Microbial Biotechnology Laboratory, School of Biosciences and Technology, VIT University, Vellore, Tamil Nadu, India*

## ABSTRACT

The Indian plastic market is facing enormous challenges from industries, technology and environmental organisations. The surging increase in the waste generated calls for newer biotechnological waste management techniques. There is an exponential increase in the usage of plastic commodities particularly low-density polyethylene. In the coming years the consumption of LDPE is expected to escalate further. There have been various conventional methods followed for recycling of plastic solid waste. The reuse of plastics will alleviate the problem caused by plastics in the environment one such successful example is usage of recycled LDPE for greenhouse application having excellent mechanical and physical properties and resistant to weathering. This book chapter focuses on the different types of low-density polyethylene, microorganism mediated degradation, different changes in the physiological properties of LDPE post degradation and its application in other fields. Microbial mediated

---

[*] Corresponding Author's E-mail: jayanthi.abraham@gmail.com.

degradation of LDPE has led the scientist to investigate the different metabolic pathways involved in the degradation process. The microbial mediated degradation of LDPE is initiated by the secretion of extracellular or intracellular enzymes thereby resulting in the cleavage of long polymer chain compounds to monomers. Also, a detailed explanation on the effect of fungal activity on LDPE is provided in this chapter.

**Keywords:** LDPE, microorganisms, enzymes, hydrophobicity, oxo-degradable LDPEs

# 1. INTRODUCTION

Man-made long chain polymeric units are known as plastics (Scott, 1990). Nowadays synthetic polymers have replaced natural materials almost in every area. They have become a requisite part of our life. Their stability and durability have been enhanced considerably, which has made them more resistant to many environmental influences. Plastics are made from inorganic and organic raw materials such as carbon, silicon, hydrogen, nitrogen, and oxygen and the other basic materials are extracted from oil, coal and natural gas. Plastic industry has gone through major revolutionary technological developments and advancements thus enhancing small and medium scale industries. Even though it has been proven to be detrimental to the environment there is no other product with a wide variety of applications catering to the growing domestic and industrial demands. The use of this polymer is becoming indispensable in emerging technologies of various fields such as, food, health care, cosmetic, construction, automobile, energy and domestic purposes. The products manufactured and used in daily life are electric goods, plastic furniture, water pipes, bottles, jars etc. as depicted in Figure 1.

Around 30% of the total plastics used worldwide are only for packaging and which amounts to a high rate of 12% per annum. The usage of paper and other cellulose-based products for packaging has decreased over the years because of certain drawbacks in terms of physical and chemical properties such as strength, resistance, lightness compared to plastics.

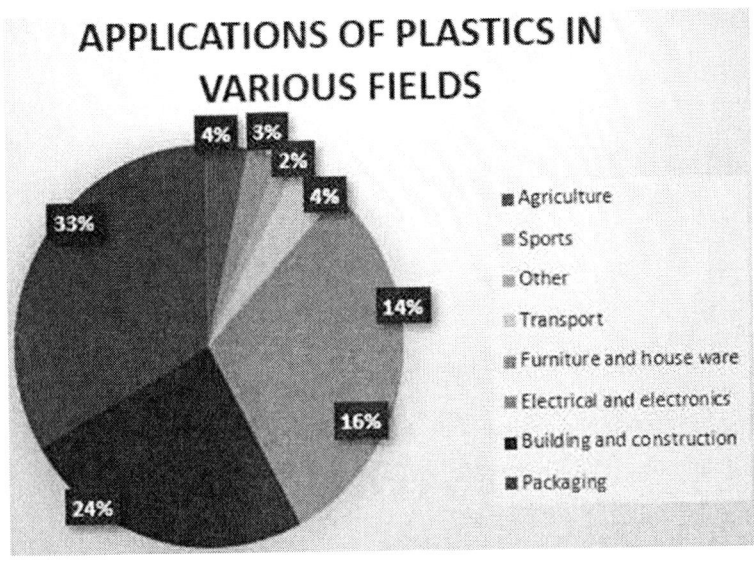

Figure 1. Applications of plastic in various fields.

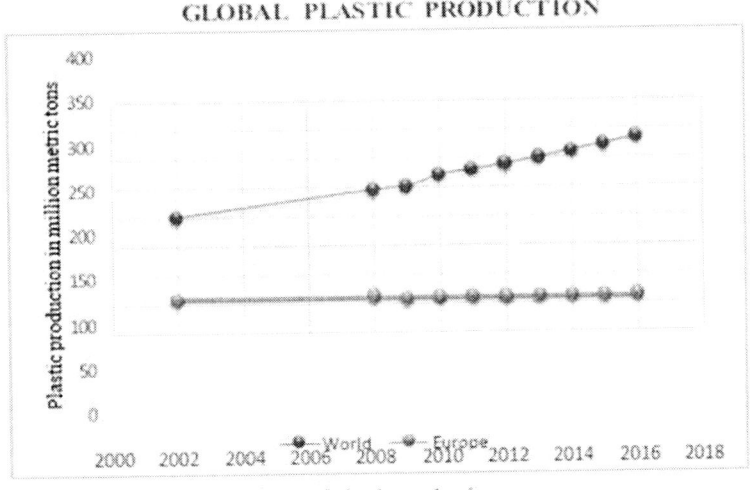

Figure 2. Global Plastic Production from 2002-2016.

It is estimated that around 147 million metric tons of petroleum-based synthetic polymers are being produced worldwide every year and bulk of these synthetic polymers being introduced into the environment are gradually increasing in the form of industrial waste products (Shimao, 2001). Global plastic production has been depicted in Figure 2.

The above statistical data portrays the global plastic production from 1950 to 2016. In 2016, world plastics production was approximately 335 million metric tons out of which 60 million metric tons of plastic production was generated in Europe alone.

Recent studies have shown that nearly 308,000 tons of plastic is being consumed only for commercial food packaging in India (Veethahavya et al., 2016) and is estimated that the usage of plastic in India is expected to increase by 129% by 2023. In the year 1999-2000, almost 120,000 tons of plastic were imported from India. 5.6 million metric tons of plastic waste is being generated in India every year of which Delhi alone accounts for an appalling amount of 689.5 metric tons of plastic waste per day.

A survey by Central Pollution Control Board (CPCB) of India found that around 9,204 tons of plastic waste is being collected every day from across the country for recycling which is roughly 60% of the total plastic waste and the remaining 6,137 tons remain uncollected and scattered. If this scenario continues the accumulation of plastics in the environment increases if left unchecked leads to impediment in pollution control and in turn affects all natural habitat and life forms.

## 2. TYPES OF PLASTIC

Plastics are broadly categorized into two types one is thermoplastics and the other thermosetting plastics. Plastics that are not affected by temperature which can be reset, remolded into desired shapes without bringing about changes in their chemical nature. They are known for the ability to be remolded repeatedly. Thermosetting plastics have many repeating molecular units which are known as repeating units. These repeating units are derived from monomers. Each polymer chain in thermosetting plastics will have several thousands of repeating units. Thermosets can be melted and molded into various desired shapes but once they are hardened, they remain solid and cannot be remolded as the chemical composition cannot be changed further. The chemical reaction that takes place during the thermosetting process is irreversible whereas the chemical reaction that takes place by heating thermoplastics is reversible. Vulcanization of rubber is a perfect example which explains the thermosetting process. The polyisoprene is a viscous material which flows readily prior to heating with sulfur, but after vulcanization, the product becomes stiff and cannot be blend thereafter.

Polyethylene (LLDPE, LDPE, MDPE, and HDPE), polystyrene (PS), polypropylene (PP), polyurethane (PUR), polyvinyl chloride (PVC), poly (ethylene terephthalate) (PET), poly (butylene terephthalate) (PBT), nylons are the most widely used plastics in packaging. These widespread applications and types of plastics are developed due to their favorable mechanical, thermal, physical and chemical properties (Rivard et al., 1995).

The commonly used plastics are:

- Polyamides: this type of plastics are generally made up of nylon which is a thick and soft plastic used for making bristles of tooth and gum brush, thin fibers and spare parts or delicate parts of automobiles.
- Polycarbonate (PC) – these are the type of plastic which is used to make shining or break proof articles like spectacular lenses, safety windows, and doors. This plastic is very transparent and hard.
- Polyester (PES) – these types of plastics are used in clothing and textile industries.
- Polyethylene (PE) – these plastics are used for packing purpose and are easy to degrade.

## 3. POLYETHYLENE (PE)

Globally, usage of plastic is growing at a rate of 12% every year. Among all the types of commodity plastics, the commercial production of polyethylene alone is around 0.15 billion tons per year globally. Thus, polyethylene is universally known as one of the most abundant commercially produced synthetic polymers in the world. Due to the increasing rate of plastic usage in every field and the commercial advantageous properties of polyethylene, the accumulation of not only the synthetic polymers like polyethylene but also the commodity plastics accumulation rate in the environment is escalating and is estimated to be 25million tons every year and is therefore considered as a serious environmental threat (Sivan et al., 2006).

Polyethylene accounts for 64% of the total plastic production all over the world. It is a linear hydrocarbon polymer which consists of a long chain of ethylene monomers ($C_2H_4$). Silk and rubber are also polymers but they naturally exist in nature and hence they undergo decay naturally and do not pose any risk of pollution to the environment. But these man-made plastic materials such as polyethylene, polystyrene etc. are extremely non-degradable and will remain in the environment for years thereby causing pollution and affecting both terrestrial and aquatic habitats (Gaurav and Sheikh, 2014). This property of polyethylene has made them recalcitrant which is due to the existence of hydrophobic carbon backbone and the high molecular weight of the polymer. This forms the basis for rapid biodegradation of polyethylene in the solid waste management.

Increase in plastic waste disposal has resulted in finding various ways to overcome the situation of plastic pollution. Some of the methods which are already in practice are landfilling and incineration. Incineration is a process were the plastic waste is burnt entirely but causes air pollution by releasing toxic gases like dioxins, carbon monoxide, and $CO_2$ into the air which results in global warming (Swapnil et al., 2015). Hence in order to protect the environment from the problem of waste accumulation, incineration converts one kind of pollution into another. The process of landfilling is also ineffective in plastic degradation. Hence an effective method for the degradation of plastics is yet to be established.

## 4. VARIOUS TYPES OF POLYETHYLENE

Polyethylene is divided into different types based on their density and melt flow index. They are HDPE (High-density polyethylene), MDPE

(Medium density polyethylene), LDPE (Low-density polyethylene), LLDPE (Linear low-density polyethylene) and VLDPE (Very low-density polyethylene). Different types of polyethylene are tabulated (Table 1) below based on their density and melt flow index.

### Table 1. Types of polyethylene (PE)

| Type of polyethylene (PE) | Density (g/cm$^3$) | Melt flow index (g/10 min) |
|---|---|---|
| HDPE | 0.941-0.965 | 0.2-3.0 |
| MDPE | 0.926-0.940 | 1.0-2.0 |
| LDPE | 0.915-0.925 | 0.3-2.6 |
| LLDPE | 0915-0.925 | 0.1-10.0 |

## 5. PROPERTIES OF POLYETHYLENE (PE)

### 5.1. Mechanical Properties

Polyethylene has low strength, hardness and rigidity compared to other plastic types but they have high ductility as well as low friction and waxy texture.

### 5.2. Chemical Properties

They are nonpolar, high molecular weight saturated hydrocarbons. Polyethylene has excellent chemical resistance to strong acids and bases. They do not dissolve at room temperature and generally, however can be dissolved in aromatic hydrocarbons like toluene and xylene at elevated temperatures. Polyethylene becomes fragile upon exposing to sunlight and do not absorb water.

## 5.3. Electrical Properties

Polyethylene is a very good electrical insulator. Though it becomes electrostatically charged effortlessly, which can be reduced by the addition of graphite and carbon black.

## 5.4. Optical Properties

Optical properties of the polyethylene depend on the thickness of the film. The film with less thickness will have high transparency and the films which have high thickness will have low transparency and high opacity. HDPE films will have high thickness and high opacity compared to LDPE whereas; LDPE will have high transparency compared to HDPE.

# 6. LOW-DENSITY POLYETHYLENE (LDPE)

Low-density polyethylene (LDPE) is the most commonly used polymer in packaging industries all over the world and it is well known for its cost effectiveness, durability and a wide range of applications in various fields. LDPE accounts for 60% of the total plastics production throughout the world annually. Among various types of plastics, LDPE is one of the most predominantly found in municipal solid waste. Due to its lightweight, chemical resistance it is used predominantly for packing food, milk and agricultural products also used in packing a variety of electronic goods, packing vehicles and so on (Swapnil et al., 2015).

Generally, plastics are known for their non-degradable nature. They are very tough with a $CH_2$ backbone and thus resistant to break making them highly recalcitrant and take thousands of years to completely degrade in the environment. LDPE's polymer chains are highly branched, this ramification results in less density and more flexibility than HDPE, MDPE and other types of plastic. Research claims that LDPE with a thickness of 60μm takes no less than 300 years to get completely degraded in the environment

(Gajendiran et al., 2016). The utility of polyethylene is continuously increasing at a rate of 12% per annum and a rough estimation of 141 million tons of synthetic polymers are being produced all over the world. Among various types of polyethylene, LDPE stands as a major threat to the environment and considered as one of the major environmental pollutants (Shimao, 2001). Due to the excessive utilization of Low-density polyethylene, the amount of LDPE getting accumulated into the environment is also found to be progressively increasing and this evokes serious biological issues resulting from improper disposal.

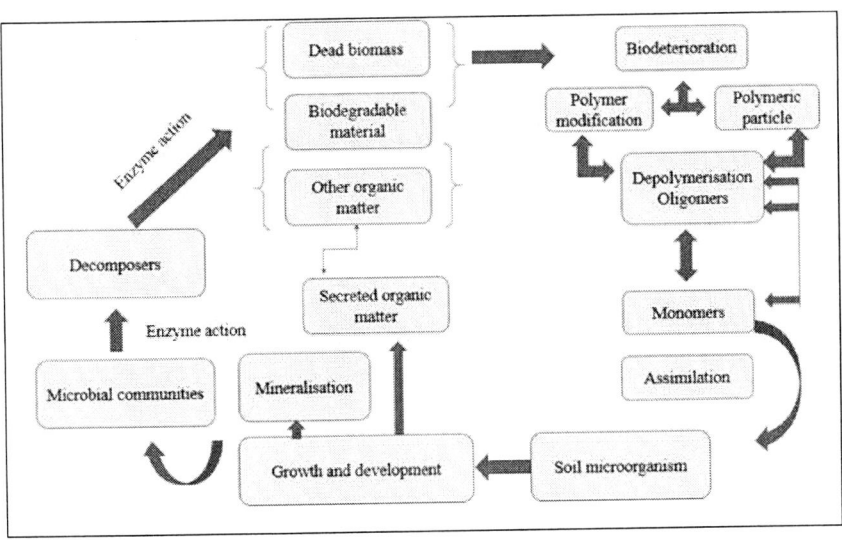

Figure 3. Depicting the overall degradation of plastic.

As polyethylene degradation takes thousands of years for complete degradation, there is a rise in production of polyethylene which is degradable in nature, carrying the same efficient features of previously produced non-biodegradable polyethylene. Overall process of biodegradation of plastic has been depicted in Figure 3.

Biodegradable plastics degraded under various environmental conditions. Depolymerization of these biodegradable plastics eventuate in the environment due to several biological factors and without human intervention to decompose these plastics and they do not pose a threat to the environment unlike artificially manufactured petrochemical plastics. The biological factors responsible for the depolymerization of these biodegradable plastics include abiotic factors like temperature, moisture, sunlight, soil present at the site, pH, salinity, humus etc. and biotic factors which comprises predominantly the microbial habitat present at the site which cause breakdown of polymers into monomer units into a non- toxic form.

## 7. DIFFERENT METHODS OF DEGRADATION

Various methods of degradation are already in practice for the efficient degradation of LDPE but none of them are efficient and every method has its own advantages and disadvantages. Chemical degradation, landfilling, biological degradation, Incineration, thermal degradation, and photodegradation are the methods of plastic degradation already in practice but unsuccessful in various levels and thus inefficient in degrading the plastics completely (Da Luz et al. 2014).

### 7.1. Thermal Degradation

In the process of thermal degradation, increased temperatures are applied for degradation of the plastic. Raised temperatures will definitely enhance the rate of chemical reactions, such as oxidation and ultimately result in the breakdown of polymers but fail to accomplish efficient degradation of polymers (Albertsson et al., 1992; Dilara and Briassoulis, 2000).

### 7.2. Photo-Degradation

The method of photodegradation is the phenomenon of de-polymerizing plastic by irradiating plastic products with UV radiation (290nm-400nm).

Absorption of UV radiation by the plastic will lead to the carbon-carbon bond cleavage hence this phenomenon of depolymerization is known as photodegradation (Dilara and Briassoulis, 2000).

## 7.3. Incineration

Incineration of plastic is the process of burning plastics. The burning of plastic wastes produces toxic gases, inhalation of these gases will pose health hazards by causing lung diseases and cancer in humans (Pramila and Vijaya, 2011). Considering inefficiency and harmful effects caused by the above methods in degrading plastic, scientists and environmentalists are focusing on safer methods for the degradation of plastic.

## 7.4. Microbial Mediated Degradation of LDPE

Biodegradation is a process mediated by microbes. Due to the disadvantages of other methods in terms of cost and pollution, using microbes for the degradation of plastic has gained more attention in recent years for the environmentally friendly disposal of plastic and polymer-based waste. Biodegradation is compatible (microbial mineralization) compared to other waste degradation management techniques (Schink et al., 1992).

Degradation of plastics mediated by microbes is broadly accepted and research is underway to enhance the efficiency by understanding molecular level metabolism by microbes involved in degradation of plastics. Metabolites involved in degradation are still yet to be deduced. Recently several microorganisms isolated from plastic polluted sites reported to produce plastic degrading enzymes. The biodegradation of plastic is achieved by the enzymatic activities shown by the microorganisms. The enzymatic activity of microbes will lead to the cleavage of the polymer into oligomers and monomers. These products formed by the breakdown of polymers are water soluble and further utilized by microbes as a carbon source.

## Table 2. Showing the different polymers degraded using microorganism

| S.No. | Types of polymer | Microorganisms involved | References |
|---|---|---|---|
| 1. | Polyethylene | *Brevibacillus borstelensis, Comamonas acidovorans* TB-35, *Pseudomonas chlororaphis, P. aeruginosa, P. fluorescens, Rhodococcus erythropolis, R. rubber, R. rhodochrous, Staphylococcus cohnii, S. epidermidis, S. xylosus, Streptomyces badius, S. setonii, S. viridosporus, Bacillus amyloliquefaciens, B. brevis, B. cereus, B. circulans, B. circulans, B. halodenitrificans, B. mycoides, B. pumilus, B. sphaericus, B. thuringiensis, Arthrobacter paraffineus, A. viscosus* | Shah et al., 2008; Bhardwaj et al., 2012; Dussud and Ghiglione, 2014; Restrepo-Flórez et al., 2014; Grover et al., 2015; Kale et al., 2015 |
|  |  | *Acinetobacter baumannii, Microbacterium paraoxydans, Nocardia asteroides, Micrococcus luteus, M. lylae, Lysinibacillus xylanilyticus, Aspergillus niger, A. versicolor, A. flavus, Cladosporium cladosporioides, Fusarium redolens, Fusarium* spp. AF4, *Penicillium simplicissimum* YK, *P. simplicissimum, P. pinophilum, P. frequentans, Phanerochaete chrysosporium, Verticillium lecanii, Glioclodium virens, Mucor circinelloides, Acremonium kiliense, Phanerochaete chrysosporium* | Shah et al., 2008; Bhardwaj et al., 2012; Dussud and Ghiglione, 2014; Restrepo-Flórez et al., 2014; Grover et al., 2015; Kale et al., 2015 |

| S.No. | Types of polymer | Microorganisms involved | References |
|---|---|---|---|
| 2. | Polyvinyl chloride | *Pseudomonas fluorescens* B-22, *P. putida* AJ, *P. chlororaphis*, *Ochrobactrum* TD, *Aspergillus niger* | Shah et al., 2008; Dussud and Ghiglione, 2014; Kale et al., 2015 |
| 3. | Polyurethane | *Comamonas acidovorans* TB-35, *Curvularia senegalensis*, *Fusarium solani*, *Aureobasidium pullulans*, *Cladosporium* sp., *Trichoderma* DIA-T spp., *Trichoderma* sp., *Pestalotiopsis microspora* | |
| 4. | Poly(3- hydroxybutyrate) | *Pseudomonas lemoignei*, *Alcaligenes faecalis*, *Schlegelella thermodepolymerans*, *Aspergillus fumigatus*, *Penicillium* spp., *Penicillium funiculosum* | Shah et al., 2008; Bhardwaj et al., 2012; Dussud and Ghiglione, 2014; Kale et al., 2015 |
| 5. | Poly(3-hydroxybutyrate co-3-hydroxyvalerate | *Clostridium botulinum*, *C. acetobutylicum*, *Streptomyces* sp. SNG9 | |
| 6. | Polycaprolactone | *Bacillus brevis*, *Clostridium botulinum*, *C. acetobutylicum*, *Amycolatopsis* sp., *Fusarium solani*, *Aspergillus flavus* | |
| 7. | Polylactic acid | *Penicillium roqueforti*, *Amycolatopsis* sp., *Bacillus brevis*, *Rhizopus delemar* | Shah et al.,2008 |

Degradation of polymers is a process that changes the strength and color of the polymer under controlled conditions. Disruption of length of the chain initiates the primary breakdown into simpler forms also various external factors such as temperature and chemicals also enhance the rate of degradation. It is utilized in polymer recycling, which reduces pollution load exhibited by the byproducts (Kumar et al., 2011; Bhardwaj et al., 2012). Anaerobic degradation is another way to degrade plastic materials through landfilling (Schink et al., 1992; Shah et al., 2008). Currently, the recycling process is increasing but the recycling rate is very low for most plastic polymers because of the usage of more additives in the manufacturing of plastics (Song et al., 1998). The recycling rate of thermosets is very low as compared to thermoplastics (Moore, 2008). Plastic comprises 60–80% of litter in the environment, and its persistence ability in the surroundings create harmful effects on wildlife as well as in the agriculture and forest land. Furans and dioxins are the most persistent organic pollutants (POPs) and are formed through the burning of polyvinyl chloride (PVC).

Plastic waste may come from post consumption and different stages of commercial product production (Shah et al., 2008; Nerland et al., 2014). Worldwide research for the past 3 decades has been focused on the biodegradation of plastic more efficiently (Shimao, 2001).

Biological agents, both prokaryotic (bacteria) and eukaryotic (fungi, algae and plant), are involved in the bioremediation process. *Streptomyces, Arthrobacter, Corynebacterium, Pseudomonas, Micrococcus* and *Rhodococcus* are the prominent microbial communities being used for bioremediation as illustrated in Table 2 (Kathiresan, 2003; Shah et al., 2008; Bhardwaj et al., 2012; Bhatnagar and Kumari, 2013; Dussud and Ghiglione, 2014; Restrepo-Flórez et al., 2014; Grover et al., 2015; Kale et al., 2015).

## 8. DIFFERENT STEPS IN THE POLYMER BIODEGRADATION MECHANISM

Microorganisms break down the complex compounds into a simpler form through biochemical pathways. Biodegradation of polymer is described as the modification of the polymer properties such as reduction in the molecular weight of the polymer, and loss of mechanical strength and surface properties, using microbial enzymes. Degraded particles are probably non-toxic to the environment and do not persist in the environment for a longer period. In nature microorganisms form catalytic enzymes for the process of biodegradation (Hadad et al., 2005). This approach is suitable for environmental waste management, and microorganisms involved in this process for oxidation serve as an alternative mode to maintain a clean environment (Singh and Sharma, 2008). The microbial degradation goes through steps such as: bio-deterioration (altering the chemical and physical properties of the polymer), bio-fragmentation (polymer breakdown in a simpler form using enzymatic cleavage) and assimilation (uptake of molecules by microorganisms) and mineralization (production of oxidized metabolites ($CO_2$, $CH_4$, $H_2O$) post degradation) as shown in Figure 4. Mineralization of polymers into simpler compounds takes place in both aerobic and anaerobic conditions. In the aerobic condition, products formed are $CO_2$ and $H_2O$ whereas under anaerobic conditions, $CH_4$, $CO_2$ and $H_2O$ are the end products (Singh and Sharma 2008). The biodegradation procedure of a few polymers is known and others are yet to be studied (Shimao, 2001).

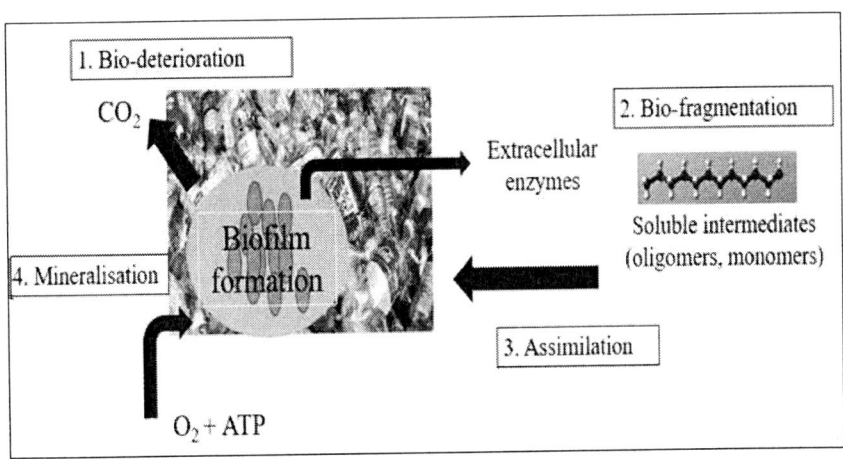

Figure 4. Pictorial depiction of different stages in polymer degradation.

Most of the microbial communities are able to utilize polyester and polyurethane at a slower rate as their sole carbon source (Schink et al., 1992; Dey et al., 2012). Starch- or cellulose based plastics are easily biodegradable; through composting, which can reduce landfilling and hence solves the waste management problem. Biodegradation using microorganisms is the best suited way to clean up plastic waste. Microorganisms are able to utilize synthetic polymers as sole carbon source, but the composition of the polymer and manufacturing process needs to be defined for the biological activity on the polymer material (Song et al., 2009;

Leja and Lewandowicz, 2010; Kumar et al., 2011; Sivan, 2011). Biodegradability of synthetic polymers with chemical groups that are susceptible to microbial attack can be done with polycaprolactone, poly-β-hydroxyalkanoates and oil-based polymers (Song et al., 2009; Leja and Lewandowicz, 2010).

Enzymes of microbial origin are employed to control pollution of xenobiotic compounds and contribute to developing an eco-friendly environment in the future. Diverse forms of microorganisms are known to utilize these polymers through the process of mineralization.

## 9. MICROBIAL METABOLISM AND PHYSIOLOGICAL PROCESSING OF POLYMER DEGRADATION

Bacteria and fungi are the widely distributed group of microorganisms that play a pivotal role in the degradation of polymer compounds in the environment (Upreti and Srivastava, 2003). These microorganisms aid in converting the insoluble biopolymer into a soluble form. Naturally occurring polymers consist of compounds formed of lipids, carbohydrates and proteins. Microbial enzymes are the ultimate source to hydrolyze low molecular weight polymers and soluble macromolecules. These soluble compounds are used by microorganisms for energy production (Gallert and Winter, 2005). Microbial degradation of polymers leads to alteration of the physicochemical properties of polyethylene compounds. The bioconversion or degradation of biomaterials is well understood by studying the mechanical properties, degradation kinetics and recognition of the degraded products or metabolites formed. The bioconversion process alters the efficacy of the host response, cellular growth, material function, etc. (Azevedo and Reis, 2005).

The tricarboxylic acid (TCA) cycle assists as one of the main metabolic pathways for energy generation from most of the organic compounds present. In the TCA cycle, acetyl-CoA acts as the key intermediate and is subjugated in cellular activities like $CO_2$ formation by oxidation, acetate formation, biosynthesis, etc. The major contribution of the TCA cycle is to generate more ATP to provide energy to the cell. Two molecules of ATP are synthesized by oxidation of 2 mol of acetate, while 34 mol of ATP is synthesized by the electron transport chain (ETC) through phosphorylation. The metabolism and efficiency of energy production differ according to the microbial growth conditions (Azevedo and Reis, 2005). Aerobic bacteria are able to respire carbohydrates, but one-third of the starting energy is not utilized by the cell and is lost in the form of heat; the remaining energy is conserved biochemically. During the processing of wastewater, activated sludge reactors they lose much of their energy as heat. Under growth-limiting conditions, the ATP consumption rate is increased and less energy is available for cellular growth and metabolism (Gallert and Winter, 2005). Some of the bacterial and fungal genera used in biodegradation of plastics have been tabulated in Table 3 and 4 respectively.

Bode et al., (2000) investigated the physical and chemical process of synthetic poly (cis-1,4-isoprene) polymer biodegradation and found that two bacterial strains, i.e., *Streptomyces coelicolor* 1A and *Pseudomonas citronellolis*, were able to utilize degraded vulcanized natural rubber and synthetic poly(cis-1,4-isoprene) as sole source of carbon for their metabolic pathway. They observed the growth of these bacteria on polymer was better than earlier reported actinomycetes *Streptomyces lividans* 1326 under controlled conditions. Three degraded products were obtained from the culture suspension of *S. coelicolor* 1A identified as (5Z,9Z)-6,10- dimethyl-penta-dec-5,9-diene- 2,13-dione, (5Z)-6-methyl-undec-5-ene-2,9-dione and (6Z)-2,6-dimethyl-10-oxo-undec-6-enoic acid. They also proposed the oxidative pathway for conversion of poly(cis-1,4-isoprene) into methyl-branched diketones by following different steps, i.e., aldehyde intermediate to carboxylic acid oxidation, β-oxidation, oxidation of the conjugated bond (double bond) to β-keto acid and the decarboxylation process. The authors proposed a hypothetical model for the degradation of poly (cis-1, 4-isoprene).

## Table 3. Different species of bacteria involved in the biodegradation of LDPE

| S.no. | Genus | Species | References |
|---|---|---|---|
| 1. | Pseudomonas | sp. | Balasubramanian et al., 2010; Tribedi and Sil, 2013 |
| | | aeruginosa | Koutny et al., 2009; Rajandas et al., 2012 |
| | | fluorescens | Nowak et al., 2011 |
| 2. | Paenibacillus | macerans | |
| 3. | Rahnella | aquatilis | |
| 4. | Ralstonia | sp. | Koutny et al., 2009 |
| 5. | Rhodococcus | erythropolis | |
| | | rhodochrous | Fontanella et al., 2010 |
| | | ruber | Santo et al., 2012 |
| 6. | Staphylococcus | cohnii | Koutny et al., 2009 |
| | | epidermidis | Chaterjee et al., 2010 |
| | | xylosus | Koutny et al., 2009 |
| 7. | Stenotrophomonas | sp. | Koutny et al., 2009 |
| 8. | Streptomyces | badius | Pometto et al., 1992 |
| | | setonii | |
| | | viridosporus | |
| 9. | Bacillus | amyloliquefaciens | Koutny et al., 2009 |
| | | brevies | Watanabe et al., 2009 |
| | | cereus | Koutny et al., 2009 |
| | | circulans | Watanabe et al., 2009 |
| | | halodenitrificans | Fontanella et al., 2010 |
| | | mycoides | Koutny et al., 2009 |
| | | pumilus | Koutny et al., 2009 |
| | | sphericus | |
| | | thuringiensis | Koutny et al., 2009 |
| 10. | Brevibacillus | borstelensis | Fontanella et al., 2010 |
| 11. | Delftia | acidovorans | Koutny et al., 2009 |
| 12. | Flavobacterium | sp. | |
| 13. | Micrococcus | luteous | Koutny et al., 2009 |
| | | lylae | |
| 14. | Microbacterium | paraoxydans | Rajandas et al., 2012 |
| 15. | Nocardia | asteroides | Fontanella et al., 2010 |
| 16. | Acinetobacter | baumannii | Koutny et al., 2009 |
| 17. | Arthrobacter | sp. | Balasubramanian et al., 2010 |
| | | paraffineus | Fontanella et al., 2010 |
| | | viscosus | Koutny et al., 2009 |

## Table 4. Different species of fungus involved in the biodegradation of LDPE

| S.no. | Genus | Species | References |
|---|---|---|---|
| 1. | Aspergillus | Niger | Koutny et al., 2009 |
| | | Versicolor | Nowak et al., 2011 |
| | | flavus | Watanabe et al., 2009 |
| 2. | Cladosporium | cladosporioides | Nowak et al., 2011 |
| 3. | Chaetomium | sp. | Watanabe et al., 2009 |
| 4. | Fusarium | Fusarium | Nowak et al., 2011 |
| 5. | Gliocladium | virens | Koutny et al., 2009 |
| 6. | Mucor | circinelloides | Balasubramanian et al., 2010 |
| 7. | Penicillium | simplicissimum | Watanabe et al., 2009 |
| | | pinophilum | |
| | | frequentans | |
| 8. | Phanerochaete | chrysosporium | Rajandas et al., 2012 |
| 9. | Verticillium | lecanii | |

Similarly, Mooney et al., (2006) studied the microbial degradation of styrene which is a potent toxic pollutant to the environment as well as human beings. It is formed during industrial processes that involve polymer and petrochemical processing. Mooney et al., (2006) reported bacterial enzymes involved in styrene biodegradation. Pyruvate, 2-vinyl muconate and acetaldehyde 2- phenylethanol are some of the degradative metabolites obtained during styrene biodegradation. Similarly, phenyl acetyl-CoA obtained via styrene degradation enters into the tricarboxylic acid (TCA) cycle. The TCA cycle thus plays a vital role in the production of energy essential for both cellular and metabolic events and thereby producing $CO_2$ during the oxidation process.

## 10. MICROBIAL DEVELOPMENT AS BIOFILMS ON POLYMER

Biofilm formation is a highly coordinated and intricate process. Some microorganisms are reported to change their morphology (e.g., *Bacillus subtilis*, *Caulobacter crescentus*) under adverse conditions, which allows them to exist in a hostile environment (O'Toole et al., 2000).

Microorganisms can grow on a wide variety of surfaces, i.e., inert or living. Inert surfaces include piping systems (biofilms on noncellular materials), medical devices (biofilms composed of microorganisms with blood components) and living tissues, which serve as surfaces for the biofilm development and consisting of single or mixed populations of microorganisms. Single species of gram-positive (*Staphylococcus aureus*, *Staphylococcus epidermidis* and *Enterococcus*) and gram-negative (*E. coli*, *Pseudomonas aeruginosa*, *Pseudomonas fluorescens* and *Vibrio cholerae*) bacteria were widely reported for the formation of biofilm. However, different species populations of microorganisms have been reported predominantly in biofilms (O'Toole et al., 2000; Prakash et al., 2003). Microorganisms that form biofilms show elaborate growth by increasing their resistance against antibiotics via gene regulation. Microorganisms reflect unique inhabitancy in biofilms with enhanced defense mechanisms against therapeutic agents and drugs, and they relieve their survival under sessile conditions (osmotic stress, desiccation, UV radiation and pH changes) (Prakash et al., 2003; Bogino et al., 2013; Sharma et al., 2015).

These microorganisms are able to acclimatise and adapt to conditions, thereby undergo a transition in their nutrient uptake from the existing environmental conditions. Microorganisms in biofilms show novel phenotypic characteristics with specific mechanisms to attach to surfaces and respond to the external signals. *Escherichia coli* O517:H7 and *Myxococcus xanthus* were known to attach to the surface under nutrient-rich environments (O'Toole et al., 2000; Sharma et al., 2015). In the initial stage, microorganisms produce some proteins such as collagen, fibrin, fibronectin and laminin as a coating material, which help in cell-to-cell adhesion in biofilm formation (Bryers et al., 2006). Biofilms exist as a matrix of extracellular polymeric substance (EPS) released by them depending on the surrounding conditions; it consists of a series of biological and physical changes during matrix formation and form the three-dimensional structure (Prakash et al., 2003; Bryers et al., 2006).

In biofilms, microorganisms are differentiated compared to free-swimming organisms and form multicellular (microcolony) complex structures with interstitial voids (Sivan et al., 2006). Microorganisms are present on the upper layer of biofilms and actively dividing compared to microbes in proximity to the surface and also differ in metabolism. Interstitial water channels in the matrix help in separating the microbial colonies and facilitate diffusion of nutrients, water, gases, enzymes, waste, signals, etc., throughout the biofilms.

Heterogeneity of the biofilm matrix is one of the distinctive features that maintain the nutritional requirement of biofilms. The matrix is composed of macromolecules, polysaccharides and water, which provide heterogeneity and form enclosed exopolysaccharides in cells; thereby protecting the biofilms from the external environment (O'Toole et al., 2000; Prakash et al., 2003).

The heterogeneity in a matrix is visible in both kinds of biofilms, i.e., single species populations of microbial biofilms and mixed species populations of microbial biofilms (Prakash et al., 2003; Bogino et al., 2013). The structural components, i.e., extracellular polymeric substances, curli fimbriae, outer membrane proteins, flagella and pili are important for biofilm formation. The relative form biofilms of flagellated motile bacteria compared to non-motile bacteria and other structural components such as curli and mannose sensitive type I, IV pili (TfP) encoding genes also play an important role in cell-cell or cell-surface attachments. Type IV pili play a significant role in bacterial motility on the surface and are important for quorum sensing. OmpR is reported as the functional gene that promotes the formation of biofilm (O'Toole et al., 2000; Prakash et al., 2003; Bogino et al., 2013).

Biofilm formation is a highly regulated event. Gene algC was reported as an alginate-producing gene that enhances the formation of biofilm which is positively regulated by the sigma factor (Kokare et al., 2009). Similarly, some of the autoinducers play a significant role in the formation of biofilm and thereby having intra- and inter-species communication, e.g., AI 1, AI 2 and N- acetyl homoserine lactones (*Vibrio fischeri* or *Aliivibrio fischeri*). Such cellular communication are used in quorum sensing signaling systems, i.e., the Rh1I/Rh1R system as observed in *P. aeruginosa* and the AfeI/AfeR system in *Acidithiobacillus ferrooxidans* (Vu et al., 2009).

In addition to extracellular appendages, cellular hydrophobicity plays a vital role in surface attachment and biofilm formation. Synthetic polymers are resistant to degradation because of their hydrophobic properties and non-polar nature which prevents the absorption of water (Khoramnejadian, 2013). Usually non-polar polymers show partial degradation because of their hydrophobic surface, and thereby inhibiting the formation of biofilms. Microorganisms are easily attached to non-polar surfaces (plastics and Teflon) compared to hydrophilic materials. Non- polar polymer shows an improved surface hydrophobicity. The fimbriae structure contains hydrophobic amino acid on their ends and mycolic acid of gram-positive bacteria helps in establishing the cell attachment to the hydrophobic surface (Prakash et al., 2003; Sarjit et al., 2015). The hydrophobic nature of the surface is one of the foremost properties during the degradation of synthetic polymer and microbial attachment to the surface.

## 11. POLYETHYLENE DEGRADATION

Polyethylene is one of the most utilized forms of synthetic non-biodegradable polymers that are hydrophobic in nature (Mahalakshmi et al., 2012). In the case of polyethylene, the biodegradability directly depends on its molecular weight. Hydrocarbon having less than 620 molecular weight favors the growth of microorganisms whereas high molecular weight polyethylene is resistant to biodegradation process. The hydrophobic nature hampers bioavailability. Physical and chemical treatment enhances degradation of high molecular weight polyethylene to a certain extent. Some of the treatments include UV irradiation, thermal treatment, photo-oxidation and oxidation with nitric acid. Oxidation of polyethylene present in the polymer increases the surface hydrophilicity thereby amplify the biodegradation (Hadad et al., 2005). Chemical treatments using 0.5 M $HNO_3$ and 0.5 M NaOH on polyethylene accelerate the biodegradation process and can be completely degraded using *Pseudomonas* sp. (Nwachkwu et al.,

2010). Polyethylene and polystyrenes having ether linkage are susceptible to monooxygenases enzyme attack (Schink et al., 1992). Biodegradation of polyethylene can be enhanced by the conventional physical treatments that cause pre-ageing via heat or light exposure. Usage of hot air ovens for abiotic oxidation was successful with incorporation of polymer degrading microorganisms. Genera *Gordonia* and *Nocardia* have shown encouraging results biodegradation (Bonhomme et al., 2003).

Some genera of microorganisms are also reported for degradation of polyethene (*Bacillus, Lysinibacillus, Pseudomonas, Vibrio, Staphylococcus, Nocardia, Micrococcus, Streptomyces, Rhodococcus, Listeria, Arthrobacter, Bravibacillus, Streptococcus, Diplococcus, Chaetomium, Moraxella, Proteus, Penicillium, Serratia, Aspergillus, Phanerochaete,* and *Gliocladium*) (Koutny et al., 2006; Arutchelvi et al., 2008; Bhardwaj et al., 2012; Restrepo- Flórez et al., 2014; Grover et al., 2015). Polyethylene is utilized as the sole carbon source by microorganisms, and biofilm formation on the surface of polyethylene shows their effectiveness on the process of degradation. Biofilm formation can be improved with the addition of mineral oil (0.05%) to the medium. Non-ionic surfactants can help in catalysing polymer biodegradation by increasing the hydrophilicity of the polymer, which helps in the adhesion of microorganisms onto the polymer (Hadad et al., 2005).

Polyethylene can be degraded by following hydro- or oxo-biodegradation. The biodegradation method solely depends on the microorganisms involved in the process. Fungi (*Mucor rouxii* NRRL 1835, *A. flavus*) and strains of *Streptomyces* are also capable of degradation of starch- based polyethylene. Degraded polymers are described by pronounced modifications in surface fractures, bond scratching and other changes like color, etc., and such alterations can be examined using analytical techniques like scanning electron microscopy and FT-IR. Degraded polymer shows conformational changes in its texture that support the growth of the microbial population (Mahalakshmi et al., 2012). *Streptomyces badius* 252 and *Streptomyces setonii* 75Vi2 are some of the few reported organisms having the ability to degrade lignocelluloses whereas *Streptomyces viridosporus* T7A acts on the degradation of heat-treated plastics (Pometto et al., 1992). Abiotic factors play a vital role in increasing the surface availability for the growth of microorganisms on polymers due to processes such as photo-oxidation, physical disintegration of polymer and hydrolysis which decreases the molecular weight (Singh and Sharma, 2008). *R. ruber* degrades polyethylene by colonizing on them and forming a biofilm on the surface of the polymer (Basnett et al., 2012). Peroxidant additives are employed in polyethylene manufacturing for agriculturally used plastic, this type of polyethylene shows susceptibility to thermal and photochemical mineralization *in vitro*. In addition to UV and heat treatments, it reduces the strength of bonds like hydroxyls and carbonyls by changing their structural conformation. (Li, 2000; Feuilloley et al., 2005). *P. chrysosporium* and *Streptomyces* sp. are known for degradation of starch-blended polyethylene (Flavel et al. 2006). Similar results were reported by Psomiadou et al. for starch-blended LDPE degradation by reduction in mechanical properties (Psomiadou et al., 1997). Arvanitoyannis et al., (1998) also reported that LDPE blended with 10% (w/w) starch enhanced its biodegradation rate by changing the mechanical properties.

Microorganisms take part in degradation via modification of their metabolic functional pathways according to environmental conditions to utilize xenobiotic compounds as sole carbon source. Bioremediation process is more-affordable and eco-friendly by following novel catabolic mechanisms (Ojo, 2007). The biodegradation process is slow but this does not indicate that ingredients in plastic material and polymers are not bioactive. Polycarbonate plastics undergo leaching of bioactive compounds, bisphenol-A monomer upon salt exposure in seawater. Commercial use of several synthetic polymers made with bioactive additives monomers, which are non-stick compounds; softeners and UV stabilizers are found in nature. Their degradation rate depends on the environmental circumstances. Symphony is a type of polymeric material that is used in polyethylene formation and is easily degradable in nature (Moore, 2008; Kumar et al., 2011).

## 12. Future Prospects

Researchers have taken interest in synthesizing biodegradable polymers as a substitute to conventional non-biodegradable polymers like polyethene made up of larger amounts of petroleum products. Bio-plastics are made up of a combination of renewable biopolymers and fossil materials. Most of the microorganisms have the ability to produce an intracellular compound, poly-β-hydroxybutyric acid. Some of the renewable bioplastics are as follows cellulose, poly lactic acid (PLA), chitosan, starch, lignin and proteins. Some of the industrial and commercial sectors are bioplastics used in cosmetic packaging, transportation, agro-food packaging, sports utility, stationery, medical appliances, textile and non-woven industries, child care, construction and toy industries. Bioplastics find application in food industries in making cups, mugs, trays, plates and cutlery which are easily biodegradable as well as reusable. Biodegradable polymers gain a greater attention in the agriculture and horticulture field. Biodegradable mulching films can be used to prevent moisture loss and hence improve the crop yield. Another promising application is use of biofilms for banana bushes which helps in protecting them from dust and environmental factors. Bioplastics can also be used in pot-plant marketing. Herbal pots are a sustainable method in horticulture. Post-harvest of the plant, the film can also be composted as they are biodegradable. A large amount of electronic parts is usually made up of plastic and thereby non degradable. This invariably results in E-waste accumulation in the environment. Bioplastics can be used in castings, circuit boards, data storage devices, touch screen computer casings, loud speakers, mobile casings, keyboard elements, vacuum cleaners, etc. Main advantages of bioplastic over conventional plastics are that they produce less carbon footprint and greenhouse gas emissions, zero littering, reduced dependence on petroleum products and are easily degradable by soil microorganisms.

## Conclusion

Microorganisms are capable of degrading inorganic and organic materials, and many researchers are interested in understanding the mechanism of biodegradation by microorganism. *P. aeruginosa, P. stutzeri, S. badius, S. setonii, R. ruber, C. acidovorans, C. thermocellum* and *B. fibrisolvens* are the dominant bacterial spp. associated with polymer degradation. *P. aeruginosa* is one of the widely reported microorganisms for polymer degradation via biofilm formation with the help of alginate-like chemicals and quorum sensing signaling systems, i.e., RhlI/RhlR. Biofilm formation enhances the efficiency of degradation later followed by the mineralization (polyethylene glycols mineralization) process. *Pseudomonas aeruginosa* CA9 has been claimed to possess the ability to biodegrade LDPE, *Pseudomonas* sp. AKS2 has been reported for the formation of biofilm on LDPE and also by enhancing the microbial growth with 26% of surface hydrophobicity and 31% of hydrolytic activity. *P. stutzeri* has been proved to degrade high molecular weight (4000– 20,000) PEG, and *Streptomyces badius* 252 and *Streptomyces setonii* 75Vi2 were known to have greater efficacy against degradation of heat-treated degradable plastics. *Rhodococcus ruber* has been reported to colonize and degrade polyethylene by forming biofilm using enzymes. Gene pudA is the most important gene from *Comamonas acidovorans* TB-35 encoding the enzyme PUR esterase which is useful for degradation of polyethylene. Some of the fungus such as *A. niger, A. flavus, F. lini, P. cinnabarinus* and *M. rouxii* are commonly found in polymer degradation. *A. niger* produces acetyl xylan esterase enzyme, which in combination with endo-xylanase helps in successful and efficient degradation of xylan polymers. *A. niger* and *A. flavus* are also suitable for the rapid mineralization of medium- length monomer units. Hence, further studies on the screening of effective microbial strains are essential to minimize polymer risks in the environment. Bioplastics is a newer type of polymer which can be used in various industrial applications replacing the conventional non biodegradable synthetic plastics.

## REFERENCES

[1] Pometto, A. L., B. T. Lee, K. E. Johnson (1992), Production of an extracellular polyeth- ylene-degrading enzyme(s) by Streptomyces species, *Appl. Environ. Micro-biol.* 58 (2):731–733.

[2] Albertsson, A. C., C. Barenstedt, S. Karlsson. (1992). Increased biodegradation of a low density polyethylene (LDPE) matrix in starch-filled LDPE materials S. *J Environ Polym Degr*, 1(4): 241-245.

[3] Arutchelvi J., Sudhakar M., Arkatkar A., Doble M., Bhaduri S., Uppara P. V. (2008). Biodegradation of polyethylene and polypropylene. *Ind J Biotechnol*, 7(1):9–22.

[4] Arvanitoyannis I., Biliaderis C. G., Ogawa H, Kawasaki N. (1998). Biodegradable films made from low-density polyethylene (LDPE), rice starch and potato starch for food packaging applications: part 1. *Carbohydr Poly*, 36(2):89–104.

[5] Azevedo H. S., Reis R. L. (2005). Understanding the enzymatic degradation of biodegradable polymers and strategies to control their degradation rate Biodegradable systems in tissue engineering and regenerative medicine. CRC Press, *Boca Raton*, pp 177–201.

[6] Nowak, B., J., Paj K. M., Drozd-Bratkowicz G. Rymarz (2011). Microorganisms participating in the biodegradation of modified polyethylene films in different soils under laboratory conditions, *Int. Biodeterior. Biodegrad.* 65 (6): 958–767.

[7] Basnett P., Knowles J.C., Pishbin F., Smith C., Keshavarz T., Boccaccini A.R., Roy I. (2012). Novel biodegradable and biocompatible poly (3-hydroxyoctanoate)/bacterial cellulose composites. *Adv Eng Mater*, 14(6):330–343.

[8] Bhardwaj H., Gupta R., Tiwari A. (2012). Microbial population associated with plastic degradation. *Sci Rep* 1(2):1–4.

[9] Bhatnagar S., Kumari R. (2013). Bioremediation: a sustainable tool for environmental management-a review. *Ann Rev Res Biol*, 3(4):974–993

[10] Bode H. B., Zeeck A., Plückhahn K., Jendrossek D. (2000). Physiological and chemical investigations into microbial degradation

of synthetic poly (cis-1, 4-isoprene). *Appl Environ Microbiol,* 66(9):3680–3685.

[11] Bogino P. C., Oliva M. D. L. M., Sorroche F. G., Giordano W. (2013). The role of bacterial biofilms and surface components in plant-bacterial associations. *Int J Mol Sci,* 14(8):15838–15859.

[12] Bonhommea S., Cuerb A., Delort A. M., Lemairea J., Sancelmeb M., Scott G. (2003). Environmental biodegradation of polyethylene. *Polym Degrad Stab,* 81:441–452.

[13] Bryers J. D., Jarvis R. A., Lebo J., Prudencio A., Kyriakides T. R., Uhrich K. (2006). Biodegradation of poly (anhydride-esters) into non-steroidal anti-inflammatory drugs and their effect on Pseudomonas aeruginosa biofilms in vitro and on the foreign-body response in vivo. *Biomaterials,* 27(29):5039–5048.

[14] Da Luz, S. Albino Paes, D. Bazzolli, Marcos R., A. J. Demuner, M. Catarina, M. Kasuya (2014). Degradation of low-density polyethylene by a novel strain *Chamaeleomyces Viridis* Abiotic and Biotic Degradation of Oxo- Biodegradable Plastic Bags by *Pleurotus ostreatus* DOI: 10.1371/journal.pone.0107438.

[15] Dey U., Mondal N.K., Das K., Dutta S. (2012). An approach to polymer degradation through microbes. *J Pharm,* 2(3):385–388.

[16] Dilara, Briassoulis (2000). Degradation and Stabilization of Low-density Polyethylene Films used as Greenhouse Covering Materials. *J Agr Eng Res,* 76(4);309-321.

[17] Dussud C., Ghiglione J. F. (2014). Bacterial degradation of synthetic plastics. In CIESM Workshop Monogr (No. 46) environment. *Polym Degrad Stab,* 29: 135–154.

[18] Feuilloley P., Cesar G., Benguigui L., Grohens Y., Pillin I., Bewa H., Lefaux S., Jama M. (2005). Degradation of polyethylene designed for agricultural purposes. *J Polym Environ,* 13(4):349–355.

[19] Flavel B. S., Shapter J. G., Quinton J. S. (2006). *Nanosphere lithography using thermal evaporation of gold.* IEEE, New York, pp 578–581.

[20] Gajendiran A., Kaustubh K., Anish Mathew., Jayanthi A. (2016). Fungal mediated degradation of low-density polyethylene by a novel strain *Chamaeleomyces Viridis* JAKA1. *RJPBCS*, 7(4): 3123-3130.

[21] Gallert C., Winter J. (2005). *Bacterial metabolism in wastewater treatment systems.* Wiley-VCH, Weinheim, pp 1–48.

[22] Gaurav and Sheikh (2014). Plastic Pollution in Cities of Mumbai and Thane. J. *RHIMRJ*, (1)4:63-70.

[23] Grover A., Gupta A., Chandra S., Kumari A., Khurana S.P. (2015). Polythene and environment. *Int J Environ Sci,* 5(6):1091–1105.

[24] Rajandas, H., S. Parimannan, K. Sathasivam, M. Ravichandran, L. Su Yin (2012). A novel FTIR-ATR spectroscopy based technique for the estimation of low-density polyethylene biodegradation, *Polym. Test,* 31 (8): 1094–1099.

[25] Hadad D., Geresh S., Sivan A. (2005). Biodegradation of polyethylene by the thermophilic bacterium *Brevibacillus borstelensis. J Appl Microbiol,* 98(5):1093–1100.

[26] Kale S. K., Deshmukh A. G., Dudhare M. S., Patil V.B. (2015). Microbial degradation of plastic: a review. *J Biochem Technol,* 6(2):952–961.

[27] Kathiresan K. (2003). Polythene and plastics-degrading microbes from the mangrove soil. *Rev Biol Trop,* 51(3):629–634.

[28] Khoramnejadian S. (2013). Microbial degradation of starch based polypropylene. *J Pure Appl Microbiol,* 7(4):2857–2860.

[29] Kokare C. R., Chakraborty S., Khopade A. N., Mahadik K. R. (2009). Biofilm: importance and applications. *Ind J Biotechnol,* 8(2):159–168.

[30] Kumar A. A., Karthick K., Arumugam K. P. (2011). Biodegradable polymers and its applications. *Int J Biosci Biochem Bioinform,* 1(3):173–176.

[31] Leja K., Lewandowicz G. (2010). Polymer biodegradation and biodegradable polymers—a review. *Pol J Environ Stud,* 19(2):255–266.

[32] Li R. (2000). Environmental degradation of wood–HDPE composite. *Polym Degrad Stab,* 70(2):135–145.

[33] Koutny, M., P. Amato, M. Muchova, J. Ruzicka, A. Delort, (2009). Soil bacterial strains able to grow on the surface of oxidized polyethylene film containing prooxidant additives, *Int. Biodeterior. Biodegrad*, 63 (3): 354–357.

[34] Santo, M., R. Weitsman, A. Sivan (2012). The role of the copper-binding enzyme – laccase – in the biodegradation of polyethylene by the actinomycete *Rhodococcus ruber*, *Int. Biodeterior. Biodegrad*, 208: 1–7.

[35] Mahalakshmi V., Siddiq A., Andrew S.N. (2012). Analysis of polyethylene degrading potentials of microorganisms isolated from compost soil. *Int J Pharm Biol Arch,* 3(5):1190–1196.

[36] Mooney A., Ward P. G., O'Connor K. E. (2006). Microbial degradation of styrene: biochemistry, molecular genetics, and perspectives for biotechnological applications. *Appl Microbiol Biotechnol*, 72(1):1–10.

[37] Moore C. J. (2008). Synthetic polymers in the marine environment: a rapidly increasing, long-term threat. *Environ Res,* 108:131–139.

[38] Nerland I. L., Halsband C., Allan I., Thomas K. V. (2014). *Microplastics in marine environments: occurrence, distribution and effects* (Re.no.6754-2014). Norwegian Institute for Water Research, Oslo, pp 1–71.

[39] Nwachkwu S., Obidi O., Odocha C. (2010). Occurrence and recalcitrance of polyethylene bag waste in Nigerian soils. *Afr J Biotechnol*, 9(37):6096–6104.

[40] O'Toole G., Kaplan H. B., Kolter R. (2000). Biofilm formation as microbial development. *Annu Rev Microbiol*, 54(1):49–79.

[41] Ojo O. A. (2007). Molecular strategies of microbial adaptation to xenobiotics in natural environment. *Biotechnol Mol Biol Rev,* 2(1):1–13.

[42] Tribedi, P. A. K. (2013). Sil, Low-density polyethylene degradation by *Pseudomonas* sp. AK S2 biofilm. *Environ. Sci. Pollut. Res. Int*. 20 (6): 4146–4153.

[43] Prakash B., Veeregowda B. M., Krishnappa G. (2003). Biofilms: a survival strategy of bacteria. *Curr Sci*, 85(9):1299–1307.

[44] Psomiadou E., Arvanitoyannis I., Biliaderis C. G., Ogawa H., Kawasaki N. (1997). Biodegradable films made from low density polyethylene (LDPE), wheat starch and soluble starch for food packaging applications Part 2. *Carbohydr Polym*, 33(4):227–242.

[45] Pramila, R., and K. Vijaya Ramesh (2011). Biodegradation of low-density polyethylene (LDPE) by fungi isolated from marine water– an SEM analysis. *AJMR*, 5(28):5013-5018.

[46] Restrepo-Flórez J. M., Bassi A., Thompson M. R. (2014) Microbial degradation and deterioration of polyethylene-a review. *Int Biodeterior Biodegrad*, 88:83–90.

[47] Rivard C., Moens L., Roberts K., Brigham J., Kelley S. (1995). Starch esters as biodegradable plastics: Effects of ester group chain length and degree of substitution on anaerobic biodegradation. *Enzyme and Microbial Tech*, 17: 848-852.

[48] Chatterjee, S., B. Roy, D. Roy, R. Banerjee (2010). Enzyme-mediated biodegradation of heat-treated commercial polyethylene by Staphylococcal species, *Polym. Degrad. Stab*, 95 (2):195–200.

[49] Fontanella, S., S. Bonhomme, M. Koutny, L. Husarova, J. M. Brusson, J. P. Courdavault, Pitteri, S., G. Samuel, G. Pichon, J. Lemaire, A. Delort (2010). Comparison of the biodegradability of various polyethylene films containing pro-oxidant additives, *Polym. Degrad. Stab*. 95: 1011–1021.

[50] Sarjit A, Tan S. M., Dykes G. A. (2015). Surface modification of materials to encourage beneficial biofilm formation. *AIMS Bioeng*, 2(4):404–422.

[51] Schink B., Janssen P. H., Frings J. (1992). Microbial degradation of natural and of new synthetic polymers. *FEMS Microbiol Rev*, 103(2/4):311–316.

[52] Scott, G. (1990). Photo-biodegradable plastics: their role in the protection of the environment *Polym Degrad Stab*, 29: 135–154.

[53] Shah A. A., Hasan F., Hameed A., Ahmed S. (2008). Biological degradation of plastics: a comprehensive review. *Biotechnol Adv*, 26(3):246–265.

[54] Sharma B. K., Saha A., Rahaman L., Bhattacharjee S., Tribedi P. (2015). Silver inhibits the biofilm formation of *Pseudomonas aeruginosa*. *Adv Microbiol*, 5(10):677-680.
[55] Shimao M. (2001). Biodegradation of plastics. *Curr. Opin. Biotechnol*, 12: 242-247.
[56] Singh B., Sharma N. (2008). Mechanistic implications of plastic degradation. *Polym Degrad Stab*, 93:561–584.
[57] Sivan A. (2011). New perspectives in plastic biodegradation. *Curr Opin Biotechnol*, 22:422–426.
[58] Sivan A, Szanto M, Pavlov V. (2006). Biofilm development of the polyethylene degrading bacterium *Rhodococcus ruber*. *Appl Microbiol Biotechnol,* 73: 346–352.
[59] Song J. J., Yoon S. C., Yu S. M., Lenz R. W. (1998). Differential scanning calorimetric study of poly (3-hydroxyoctanoate) inclusions in bacterial cells. *Int J Biol Macromol*, 23:165–173.
[60] Swapnil K. Kale, Amit G. Deshmukh, Mahendra S. Dudhare, Vikram B. Patil. (2015). Microbial degradation of plastic: a review. *J Biochem Tech*, 6(1): 952-961.
[61] Watanabe, T., Y. Ohtake, H. Asabe, N. Murakami, M. Furukawa (2009). Biodegradability and degrading microbes of low-density polyethylene, *J. Appl. Polym.Sci*, 111 (1):551–559.
[62] Upreti M. C., Srivastava R. B. (2003). A potential Aspergillus species for biodegradation of polymeric materials. *Curr Sci*, 84(11):1399–1402.
[63] Balasubramanian, V., K. Natarajan, B. Hemambika, N. Ramesh, C. S. Sumathi, R. Kottaimuthu, V. Rajesh Kannan (2010). High-density polyethylene (HDPE)- degrading potential bacteria from marine ecosystem of Gulf of Mannar, India. *Lett. Appl. Microbiol*, 51 (2):205–211.

[64] Veethahavya, Vimala P. P, Dr. Lea Mathew (2016). Biodegradation of Polyethylene using *Bacillus subtilis*. *Science direct Procedia Technology*, 24:232 – 239.

[65] Vu B, Chen M., Crawford R. J., Ivanova E. P. (2009). Bacterial extracellular polysaccharides involved in biofilm formation. *Molecules*, 14(7):2535–25.

In: Low-Density Polyethylene
Editor: Johan Geisler
ISBN: 978-1-53618-192-0
© 2020 Nova Science Publishers, Inc.

*Chapter 4*

# MEMBRANE SEPARATIONS USING LOW-DENSITY POLYETHYLENE MEMBRANES

### *Alena Randová*[*], *Lidmila Bartovská and Karel Friess*
Department of Physical Chemistry,
University of Chemistry and Technology, Prague, Czech Republic

## ABSTRACT

Low-density polyethylene (LDPE) belongs to the family of thermoplastic materials. LDPE is formed from the ethylene monomer units and its density is generally low due to the branching from the main chain and due to the lower portion of solid, impermeable crystalline parts. Properties of LDPE, such as stability, resistance and toughness, make this material suitable for various applications, especially in the packaging industry (containers, bottles, laboratory equipment, bags etc.). Nevertheless LDPE sorbs some amounts of gases and liquids. Therefore, the study of LDPE sorption phenomena of various penetrants constitutes an important issue. The detailed knowledge of the preferential sorption, i.e., the sorption of one component of the mixture, together with the total sorption of material under specific conditions and penetrant compositions

---

[*] Corresponding Author's E-mail: randovaa@vscht.cz.

can reveal new information regarding LDPE properties. And, consequently, the usage of LDPE in membrane separations is promoted.

**Keywords**: sorption, separation, membrane

# LIST OF SYMBOLS

| Symbol | Property | Unit |
|---|---|---|
| $m_0$ | mass of binary solution brought in contact with 1 g of polymer | g |
| $M_1$ | molar mass of pure component 1 | g mol$^{-1}$ |
| $M_2$ | molar mass of pure component 2 | g mol$^{-1}$ |
| $m_{LP}$ | mass of the swollen polymer membrane | g |
| $m_P$ | mass of the dry membrane | g |
| $M^s$ | average molar mass of binary sorbed liquid | g mol$^{-1}$ |
| $n^s$ | number of moles of substances sorbed in 1 g of polymer | mol g$_P^{-1}$ |
| $Q^c_m$ | calculated swelling degree without crystalline phase | g g$_P^{-1}$ |
| $Q_m$ | swelling degree | g g$_P^{-1}$ |
| $V^{add}$ | volume calculated according to additivity rules | cm$^3$ g$_P^{-1}$ |
| $V^E$ | excess volume of binary liquid mixture | cm$^3$ mol$^{-1}$ |
| $V^{exp}$ | real volume | cm$^3$ g$_P^{-1}$ |
| $V_{m1}$ | molar volume of pure liquid 1 | cm$^3$ mol$^{-1}$ |
| $V_{m2}$ | molar volume of pure liquid 2 | cm$^3$ mol$^{-1}$ |
| $w_1^{ter}$ | weight fraction of component 1 | |

| Symbol | Property | Unit |
|---|---|---|
| $w_2^{ter}$ | weight fraction of component 2 | |
| $w_p^{ter}$ | weight fraction of polymer | |
| $x_1^b$ | molar fraction of component 1 in bulk liquid surrounding polymer | |
| $x_1^s$ | molar fraction of component 1 in liquid sorbed in polymer | |
| $x_2^b$ | molar fraction of component 2 in bulk liquid surrounding polymer | |
| $x_2^s$ | molar fraction of component 2 in liquid sorbed in polymer | |
| $x_i^b$ | molar fraction of component $i$ in equilibrium solution | |
| $x_{i,0}^b$ | molar fraction of component $i$ in bulk initial (index 0) solution | |
| $\beta$ | percentage relative specific volume increment | % |
| $\delta$ | solubility parameter | MPa$^{1/2}$ |
| $\rho_P$ | density of dry polymer | g cm$^3$ |
| $\Omega_2$ | preferential sorption | mol g$_P^{-1}$ |

# 1. ABOUT PE

## 1.1. Introduction

Polyethylene or polythene (PE) is the most common plastic with the chemical formula $(C_2H_4)_n$. PE is largely used in packaging, which accounts for ~ 40% of the total demand for plastic products with over a trillion plastic bags per year [1]. Its primary use is packaging (containers, bottles, bags etc.).

PE is restricted in certain applications due to its low melting point, solubility or swelling in hydrocarbons and tendency to crack when stressed [2].

Although the manufacture of PE products is simple, the disposal of PE as waste is very problematic. The problem of PE is its synthetic polymeric nature, therefore the decomposition in nature is long-standing. PE cannot be simply replaced, but the ecological route would be adapted during production to be easily degradable and environmentally friendly after use. At the same time, production must be inexpensive to make the polymer readily available.

## 1.2. History

For the first time, Polyethylene was synthesized by the German chemist Hans von Pechmann who prepared it accidentally in 1898 while investigating diazomethane [3]. A pillbox was presented in 1936 made from the first pound of polyethylene (PE). In 1933, Eric Fawcett and Reginald Gibbon investigated the high-pressure reaction of ethylene with benzaldehyde.

After this experiment which failed (benzaldehyde was obtained unchanged), it was found that there was a sub-gram amount of white waxy solid in the reaction vessel. The product was correctly identified as an ethylene polymer.

This experiment, however, was not reproducible until 1935 when Michael Perrin established a set of conditions that could be used for the consistent polymerization of ethylene. The key factor to material reproducibility was contamination of ethylene with trace amounts of oxygen. Oxygen reacted with ethylene to form peroxides, which subsequently decomposed to give free radicals that initiated the polymerization process.

Polyethylene produced by Perrin was a drawing material with a melting point of 115 °C. Nowadays, such material is known as low-density polyethylene, LDPE. In 1936, the British company Imperial Chemical Industries received the first patent for the production of PE [4].

Further, the patent of commercial production of new linear polyethylene with high density and polypropylene was granted to Phillips Petroleum, it showed that two of its scientists, Robert Banks and J. Paul Hogen, actually produced both polymers in another way in 1951. In 1953, German chemist Karl Ziegler developed a catalytic system based on titanium halides and organoaluminium compounds that operated under even milder conditions than the Phillips catalyst. However, the Phillips catalyst was cheaper and easier to work with. In the late 1950s, both Phillips and Ziegler catalysts were used to produce high-density polyethylene (HDPE). [5].

## 1.3. Types of PE and Differences in Their Application and Properties

Polyethylene has low strength, hardness and stiffness, but has high ductility and impact strength and low friction. It exhibits strong creep under permanent forces which can be reduced by adding short fibers. The commercial applicability of PE is limited by its relatively low melting point. Polyethylene consists of high molecular weight non-polar saturated hydrocarbons (chemical behaviour is close to paraffin). The individual macromolecules are not covalently linked. The macromolecules are symmetrical molecular structures, therefore they tend to crystallize. Therefore, each kind of PE is partially crystalline. The amount of the crystalline phase also determines material properties. The higher crystallinity means higher stability because the crystalline phase of PE cannot be easily dissolved at room temperature. Generally, PE exhibits also good chemical resistance because it is resistant to strong acids or strong bases. On the other hand, PE can absorb organic liquids, especially hydrocarbons, but water is absorbed only to a limited extent.

Furthermore, PE cannot be printed or glued without specific pre-treatment. High-strength joints can be easily achieved by welding plastics. It can also be easily electrostatically charged. The coal embedding/loading into PE is used as a solution to this phenomenon. PE is usually transparent and colourless material. Polyethylene is classified according to density and branching. Its mechanical properties strongly depend on properties such as the type of branching and crystal structure (Effect of crystallinity, part 5). Further, there main types of polyethylene, are:

- *Ultra-high-molecular-weight polyethylene* (UHMWPE): The high molecular weight makes it a very tough material, but results in less efficient packing of the chains into the crystal structure as evidenced by its lower density than that of high-density polyethylene. Over the years, ultra-high-molecular-weight polyethylene has emerged as the material for fabricating one of the bearing components in various arthroplasties, such as acetabular cups, tibial inserts, and glenoid sections. These components have performed in vivo. The only major concern is wear and the effect of the wear particles on the in vivo longevity of the prosthesis [6]. UHMWPE can be made through any catalyst technology, although Ziegler catalysts are most common. The applications are various, for example, bottle-handling machine parts, bearings, gears, artificial joints, and butchers' chopping boards.
- *High-density polyethylene* (HDPE): HDPE has a linear structure, with little or no branching. Short and/or long side-chain molecules exist with the polymer's long main chain molecules. The longer the main chain, the greater the number of atoms, and consequently, the greater the molecular weight. The molecular weight, the molecular weight distribution and the amount of branching determine many of the mechanical and chemical properties of the end product. High-density polyethylene resin has a greater proportion of crystalline regions than low-density polyethylene. HDPE is defined by a density of greater or equal to 0.941 g cm$^{-3}$. HDPE can be produced by chromium/silica catalysts, Ziegler-Natta catalysts or metallocene

catalysts; by choosing catalysts and reaction conditions, the small amount of branching that does occur can be controlled. HDPE is resistant to many different solvents and has a wide variety of applications, including containers, laundry detergent bottles, milk jugs, fuel tanks for vehicles, folding chairs, portable basketball system bases, chemical-resistant piping systems, geothermal heat transfer piping systems, root barrier, ballistic plates, etc. HDPE is the main polyethylene type in the Russian market. In 2007, polyethylene consumption in Russia was broken down in the following way: HDPE – 51%, LDPE – 42%, LLDPE – 7%. Russia features traditional polyethylene consumption structure: tare and package (30%), consumer and household commodities (22%), polyethylene films (19%), pipes and pipeline fittings (9%), insulation and cable insulation (8%), industrial-purpose items and other related products (10%) [7].

- *Medium-density polyethylene* (MDPE): MDPE is defined by a density range of 0.926-0.940 g cm$^{-3}$. The use of medium density polyethylene pipes for water and gas distribution is increasing. In these applications, PE structures are mainly subjected to creep loadings [8]. MDPE can be produced by chromium/silica catalysts, Ziegler–Natta catalysts, or metallocene catalysts.

- *Linear low-density polyethylene* (LLDPE): In the 70th, interest has grown greatly in linear low-density polyethylene (LLDPE) manufacturing all over the world and LLDPE has gradually replaced conventional high-pressure low-density polyetylene through its superior mechanical and thermal properties. Needless to say, LLDPE has short-chain branchings derived from comonomer units which are α-olefins, such as butene, hexene, octene, and 4-methylpentene. LLDPE is defined by a density range of 0.915-0.925 g cm$^{-3}$. LLDPE has higher tensile strength than LDPE, and it exhibits a higher impact and puncture resistance than LDPE. The applications are toys, containers, pipes [9, 10].

- *Low-density polyethylene* (LDPE): which is the subject of this chapter is discussed in details in the following parts.

## 1.4. Degradability of PE

The non-degradable plastic materials are commonly used in industry and agriculture. Due to their high resistance, they accumulate in the environment at a rate of about 25 million tons per year [11]. Polyethylene is produced from ethylene – mainly obtained from petroleum, even if it can be prepared from renewable resources. PE products have short service time, but long-existing time, therefore their presence is a burning problem for the environment. In Japan, for example, there is increasing the recycling of plastics, but it still has a large amount of plastic packaging that goes to waste. There are a lot of species of bacteria that can degrade polyethylene [11, 12, 13]. When PE is exposed to ambient sunlight, the plastic generates two greenhouse gases, methane and ethylene. Of particular concern is the type of plastic that releases gases at the highest rate: low-density polyethylene. This property makes it easier to decompose over time, thus more rapidly contaminating the environment.

# 2. ABOUT LDPE

Low-density polyethylene (LDPE) is a thermoplastic made of monomeric ethylene. It was the first polyethylene produced (free-radical polymerization), and today production is essentially the same as then. Approximately 5.7% of LDPE is recycled. Despite competition from more modern polymers, LDPE remains an important class of plastics. The high degree of long-chain branching gives LDPE unique and desirable properties.

LDPE has excellent resistance to acids or bases, alcohols, esters; quite good resistance to aldehydes, ketones; limited resistance to aliphatic and aromatic hydrocarbons and oxidizing agents; and poor resistance to halogenated hydrocarbons. LDPE is widely used for the production of various containers, dispensing bottles, washing bottles, hoses, plastic parts for computer components, playground slides, plastic wraps, and a lot of laboratory equipment. Last not least is the using in plastic bags.

LDPE density is defined within the range 0.917-0.940 g cm$^{-3}$. It is not reactive at room temperature except for strong oxidizing agents and some solvents – they cause sorption and swelling. This fact is a reason for observation of polymer + solvent systems (following parts). It continuously withstands temperatures around 70 °C.

**Table 1. The properties of LDPE**

| Property | Value | Unit | Reference |
|---|---|---|---|
| density | 0.917-0.940 | g cm$^{-3}$ | [14] |
| crystallinity | 45 | % | [15] |
| tensile yield strength | 12.4 | MPa | [16] |
| tensile rupture strength | 12.0 | MPa | [16] |
| solubility parameter | 15.8-16.8 | MPa$^{1/2}$ | [17] |
| melting enthalpy | 4.1±0.2 | kJ / mol CH$_2$ | [15] |

LDPE has a lower density compared to HDPE. Due to its more branched structure, the intermolecular forces are weaker, tensile strength is lower and elasticity is higher. To lower density of LDPE also contributes less tightly packed chains and lower portion of the crystalline phase. The other important properties of LDPE are summarized in Table 1.

# 3. MEMBRANE SEPARATION APPLICATIONS USING LDPE MEMBRANES: GAS AND LIQUID STATES AND PERVAPORATION

In 1748 the first permselective membrane was described – it was a pig bladder. In the following years, descriptions of membranes such as the bladder, parchment, nitrocellulose, live-cell packaging, etc. appear in books and journals. The most common material for making membranes today are polymers. They are high molecular weight compounds where several basic building blocks are repeated. Figure 1 shows the basic building block of PE.

$$[-CH_2-CH_2-]_n$$

Figure 1. Polyethylene.

Other industrially important polymer membranes are polytetrafluoroethylene (Teflon®), polydimethylsiloxane, polystyrene, polysulfone, polyvinylalcohol, polyvinylchloride (PVC). The proportion of the crystalline phase to some extent determines the properties of the polymer [15, 18-21]. The following table summarizes several polymers and their crystalline phase.

## 3.1. Gas Separation

Polymer membranes are already part of technological processes in several industries for the separation of gas or vapour mixtures. For gas and vapour separation, the flow of matter through the membrane is most often caused by pressure or concentration gradient. In general, the separation membrane with a high permeability usually has little selectivity. Specifically, LDPE exhibits very low water absorption. The study of LDPE in the gas medium were presented in ref. [15] (heptane, toluene).

Table 2. Crystalline phase of common polymers

| Polymer | Crystalline Phase (%) |
|---|---|
| polystyrene | 0 |
| PVC | 10 |
| Teflon® | 60-80 |
| LDPE | 45-55 |
| HDPE | 70-80 |

For example in ref. [22] researchers discussed the permeability models that exist for LDPE. They checked the effect of temperature, film thickness and pressure on the permeability of $CO_2$ and $O_2$ through LDPE. The results show that as film thickness increases, permeability and solubility increase, pressure has a minor effect on permeability in the case of $O_2$, and increasing temperature increases permeability.

## 3.2. Liquid Sorption

The sorption of a binary liquid mixture (index 1 and 2) in a polymer (index P) is characterised by two parameters: the swelling degree ($Q_m$) and the preferential sorption ($\Omega_2$) [23, 24].

On immersion into a liquid, the polymer imbibes a certain amount of liquid. This process can be quantified by the swelling degree (the relative mass increase):

$$Q_m = \frac{m_{LP} - m_P}{m_P} \tag{1}$$

where $m_{LP}$ is the mass of the swollen polymer membrane and $m_P$ is the mass of the dry membrane.

When a polymer is in contact with a binary liquid mixture, in most cases one of the mixture components is more sorbed into the polymer. The extent of this phenomenon is characterised by the preferential sorption, i.e., by the excess number of moles of certain component sorbed in the polymer compared to its number in the bulk solution having the same total number of moles of liquid mixture $n^s$ as the mixture sorbed in the polymer. If, for example, the component 2 is preferentially sorbed, the preferential sorption related to a unit of mass of the dry polymer (in mol $g_P^{-1}$) is given by the relation

$$\Omega_2 = n^s \left( x_2^s - x_2^b \right) \tag{2}$$

where $x_2^s$ is the molar fraction of component 2 in liquid sorbed in the polymer, $x_2^b$ is the molar fraction of component 2 in the bulk binary liquid surrounding the polymer. The total number of moles of substances sorbed in one gram of polymer $n^s$ can be expressed as

$$n^s = \frac{Q_m}{M^s} = \frac{Q_m}{x_2^s \cdot M_2 + x_1^s \cdot M_1} \tag{3}$$

where $M_1$ and $M_2$ are the molar masses of pure components 1 and 2, and $M^s$ is the average molar mass of binary sorbed liquid. Preferential sorption $\Omega_2$ is available from experimental data using equation

$$\Omega_2 = \frac{\dfrac{m_0}{x_{1,0}^b \cdot M_1 + x_{2,0}^b \cdot M_2}}{m_P}\left(x_{2,0}^b - x_2^b\right) \tag{4}$$

where $m_0$ is the mass (grams) of the binary solution brought in contact with $m_P$ grams of polymer, $x^b_{i,0}$ and $x^b_i$ are the molar fractions of component $i$ in a bulk initial (index 0) and equilibrium solutions. The combination of equations allows determination of the molar fraction of component in the polymer phase ($x^s_i$) [23, 24]:

$$x_2^s = \frac{Q_m \cdot x_2^b + \Omega_2 \cdot M_1}{\Omega_2 \left(M_1 - M_2\right) + Q_m} \tag{5}$$

The graph of this property as a function of bulk phase composition shows whether the membrane separation contains an azeotrope and whether the curve is distant from $x = y$ curve.

## 3.3. Pervaporation

Pervaporation is a membrane separation process for liquid separation using a polymeric membrane as the separating barrier. When a membrane is in contact with a liquid mixture, one of the components can be preferentially removed from the mixture, despite the component concentration. In order to ensure the continuous mass transport, very low absolute pressures (vacuum) are maintained at the downstream side of the membrane, or a sweeping gas can be used in the downstream side of the membrane. Generally, polymers such as PE (LDPE, HDPE) can be used for organic mixture separation or removal of organic contaminants from water [25, 26].

The phase change from liquid to vapour takes place in pervaporation. Processes involving phase changes are generally energy-intensive (like distillation). Pervaporation deals only with the minor components of the liquid mixtures and uses the most selective membranes. The first feature effectively reduces the energy consumption of the pervaporation process. Compared to the distillation, only the minor component in the feed consumes the latent heat, the overall pervaporation process runs under the constant temperature The second feature generally allows pervaporation the most efficient liquid-separating technology. The separation of benzene/cyclohexane mixture, for example, has selectivity 1.6 [25].

## 4. AZEOTROPES IN ℓ-G AND MIXTURE SEPARATION BY LDPE

Separation of the liquid mixture using a membrane is particularly important if the vapour-liquid equilibrium (VLE) contains an azeotrope. An azeotrope is a mixture of two liquids which cannot be separated by simple distillation because the vapour has the same composition as the liquid (Figure 2). Although the difference between the composition of the liquid in solution ($x_1^b$) and the liquid in the membrane ($x_1^s$) is small, if the membrane separation curve does not contain an azeotropic point, separation with the appropriate membrane is possible.

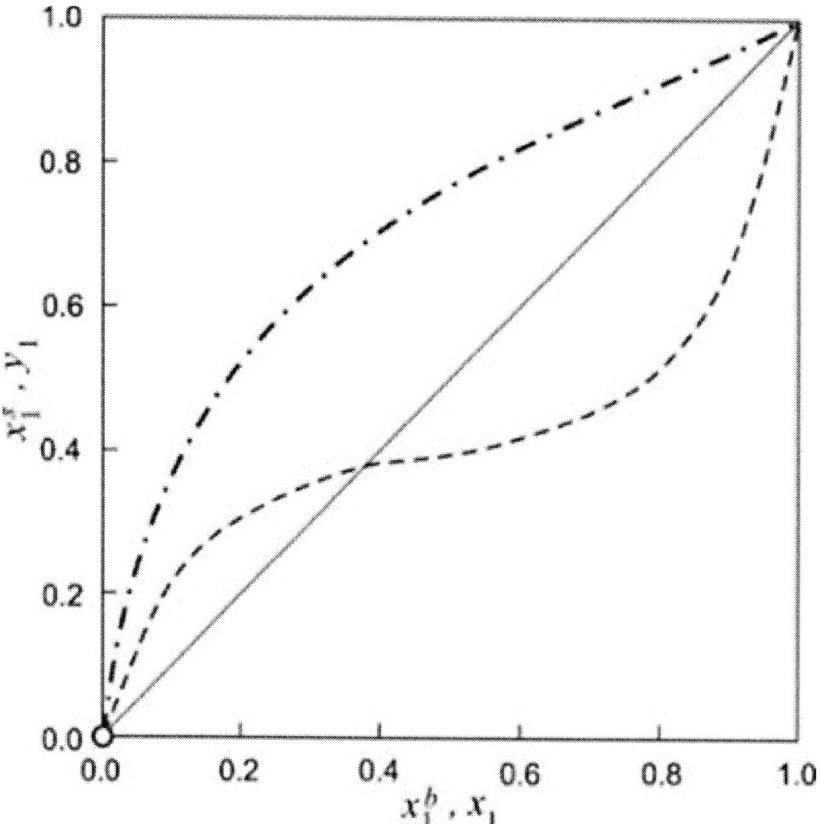

Figure 2. Separation of liquid mixture: - - - VLE, $- x_1 = y_1$, $- \cdot -$ membrane separation.

The systems with azeotropes were measured and results are presented in Figure 3. The experiments with mixture benzene + methanol (Figure 3a) were reported in ref. [23], the sorption equilibrium in the system acetone + hexane + LDPE membrane (Figure 3b) was measured for this chapter using the experimental setup presented in ref. [23].

Both mixtures have an azeotrope on VLE, but neither has an azeotrope in membrane separation, therefore LDPE could be most likely used to separate these mixtures.

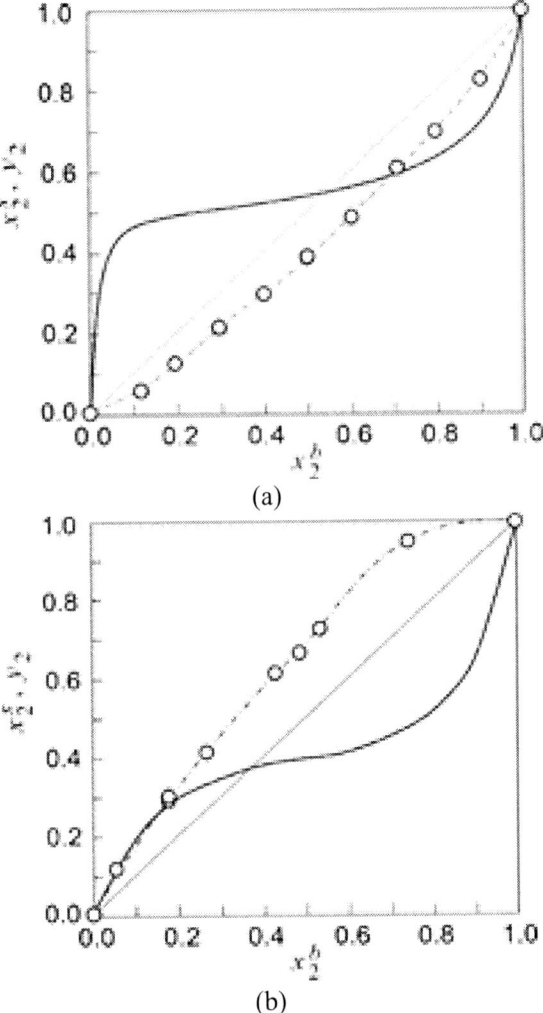

Figure 3. Equilibrium at 25 °C and atmospheric pressure (a) methanol (2) + benzene (1): – VLE, O LDPE-membrane equilibrium [23] (b) acetone (1) + hexane (2): – VLE, O LDPE-membrane equilibrium.

## 5. Effect of Crystallinity

The crystalline fraction inside the LDPE matrix specifically influences the membrane performance. It is generally assumed that the crystalline domains are not accessible to gas, vapour or liquid penetrants [15] and that all mass transport is conveyed via the amorphous phase only. The crystalline domains impose a constraint on the polymer chains in the amorphous phase and they behave as physical cross-links, restricting the swelling or sorption of a penetrant in the polymer network.

The crystallinity of LDPE is usually about 45% (Table 1 and 2).

## 6. TERNARY DIAGRAMS: WITH AND WITHOUT CRYSTALLINE PART OF THE MEMBRANE

The composition of the ternary phase (polymer + liquid component 1 + liquid component 2) is expressed in weight fractions because the molar mass of the polymer is not known [23, 27]. The weight fractions of individual components can be calculated as (the base for all the above-mentioned quantities is 1 g of dry polymer membrane):

$$w_P^{ter} = \frac{1}{1+Q_m} \tag{6}$$

$$w_1^{ter} = \frac{n_1^s \cdot M_1}{1+Q_m} \tag{7}$$

$$w_2^{ter} = 1 - w_1^{ter} - w_P^{ter} \tag{8}$$

The composition of ternary phase and composition of bulk binary liquid shows the graph in Figure 4.

Because is the assumption that the liquid sorbs only into the amorphous phase, not into the crystalline domains, the examples (Figure 5) of ternary diagrams are presented with (a) and without (b) crystalline phase.

System LDPE + butan-1-ol + heptane was presented in ref. [27], system LDPE + benzene + methanol was presented in ref. [23]. The data without crystalline phase were calculated from these refs. according to the following equation:

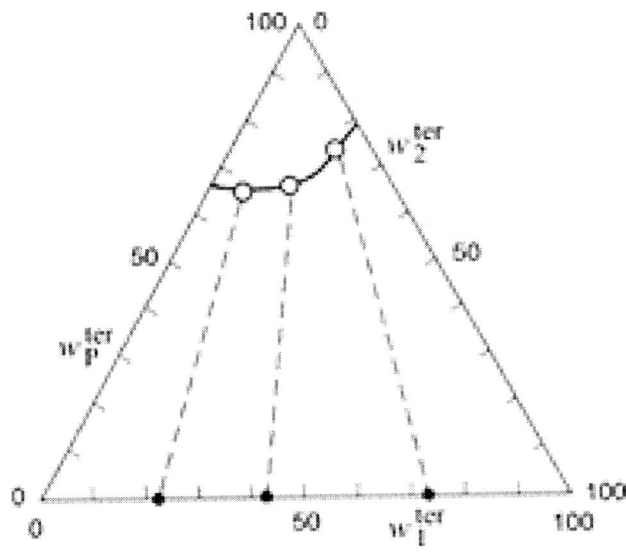

Figure 4. Ternary diagram at 25 °C and atmospheric pressure: ○ the composition of ternary phase, ● the composition of the bulk binary liquid, - - connecting lines of respective compositions.

$$Q_m^c = \frac{Q_m}{1-0.45} \qquad (9)$$

where $Q^c{}_m$ is a calculated swelling degree without crystalline phase.

This figure illustrates the big difference in the calculation when the crystalline phase is taken into account with and when not. Although it would appear that sorption in a polymer that does not contain a crystalline phase at all and sorption in a crystalline phase polymer where only the crystalline phase is subtracted is the same, it is not so [15, part 5].

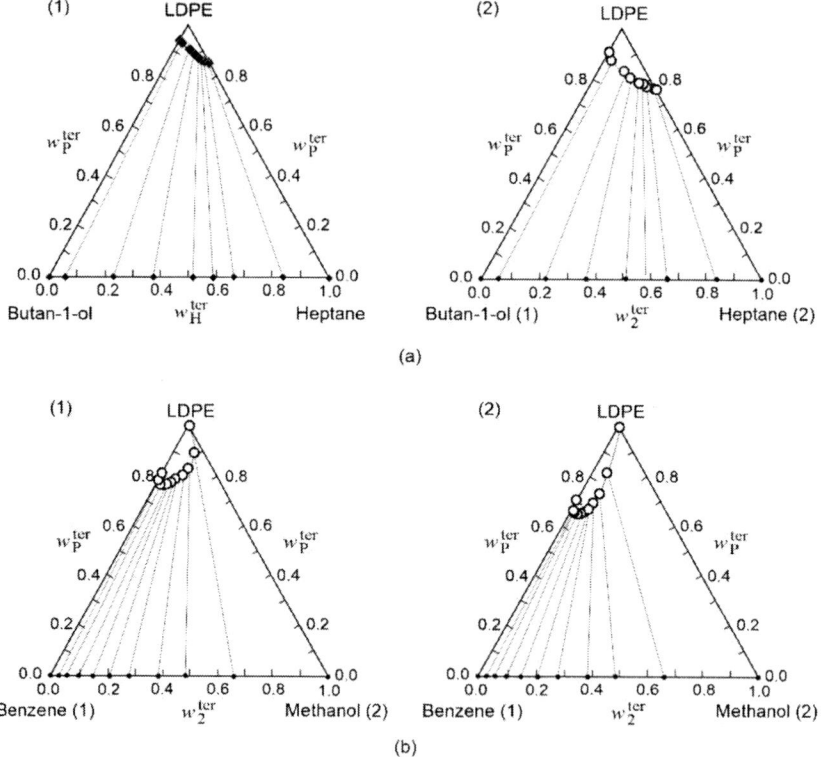

Figure 5. Ternary diagrams: (a) LDPE + butan-1-ol + heptane, (b) LDPE + benzene + methanol (1) with crystalline phase, (2) without crystalline phase.

## 7. VOLUME OF SWOLLEN MEMBRANE

The volume changes of the membrane in contact with the liquid medium are often attributed only to the sorbed liquid [23, 27-31]. The specific volume of the swollen membrane is then calculated under the assumption of additivity as

$$V^{add} = \frac{1}{\rho_P} + n^s \left( x_1^s \cdot V_{m1} + x_2^s \cdot V_{m2} + V^E \right) \tag{10}$$

where $\rho_P$ is the density of the dry polymer, $n^s$ the equilibrium mole number of liquid sorbed into 1 g of the dry membrane, $V_{m1}$ and $V_{m2}$ the molar volumes of pure liquids, and $V^E$ the excess volume of a binary liquid mixture. The examples of comparison of real volume ($V^{exp}$) and volume calculated according to additivity rules ($V^{add}$) are presented in the following figure.

In numerous cases the real volume of the swollen membrane ($V^{exp}$) is different from $V^{add}$ calculated as the simple sum of polymer and liquid volumes, i.e., assuming zero mixing volume of polymer with a binary liquid solution. This difference can be attributed to the interactions between the polymer and the liquid medium. The percentage relative specific volume increment as a function of the solution, the composition is shown in Figure 7.

$$\beta = \left( \frac{V^{exp}}{V^{add}} - 1 \right) \cdot 100 \tag{11}$$

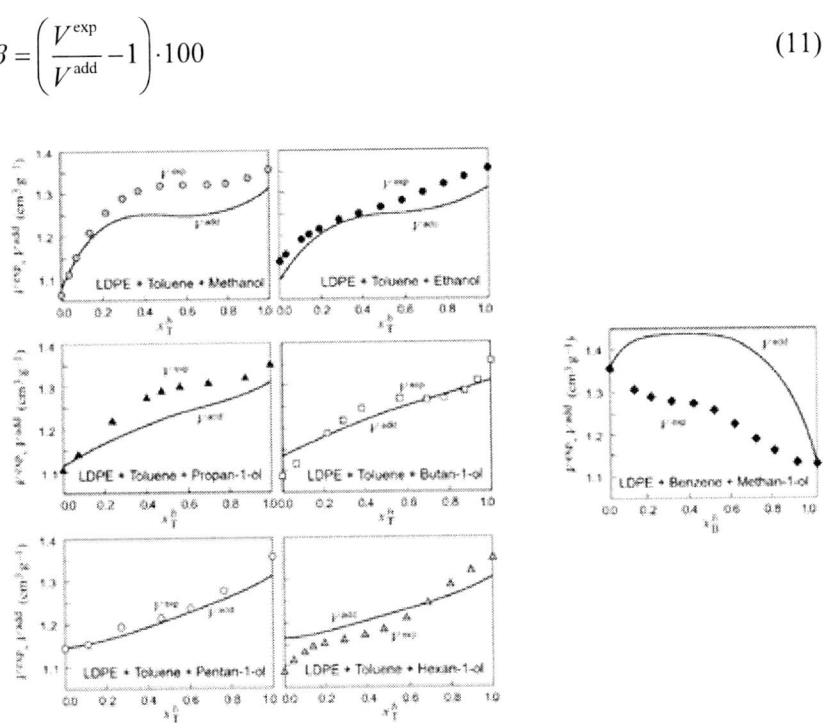

Figure 6. The comparison of the experimental volume of the swollen membrane ($V^{exp}$-points, $V^{add}$-lines): (up) LDPE + toluene + alcohol at 25 °C [29] (down) LDPE + benzene + methanol at 25 °C [23].

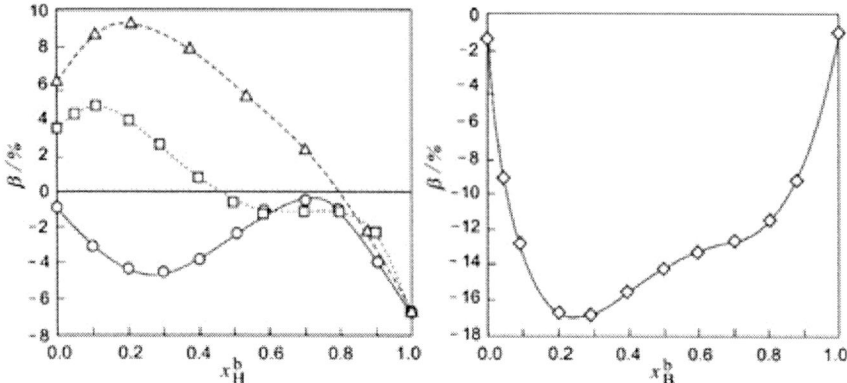

Figure 7. Percentage relative specific volume increment $\beta$ in the mixture: (left) LDPE + hexane (H) + benzene (circles), methylbenzene (squares), and ethylbenzene (triangles) at 25 °C [28] (right) LDPE + benzene (B) + methanol at 25 °C [23].

It can be seen that the percentage relative volume increment is not negligible, therefore real volume cannot be calculated from mass changes, it must be measured.

## CONCLUSION AND FUTURE OF LDPE

In today's global economic situation, scientists and engineers are forced to respond to the rapidly changing needs of society and industry. In general, membrane separation processes can better protect the environment than energy-intensive conventional processes (such as distillation). Limited natural resources must also be saved. One of the readily available and inexpensive membranes is LDPE. This membrane is tested in both gaseous and liquid separation processes. For example, it is tested for mixtures that have an azeotrope on the x-y curve, so distillation is not suitable for them. Even with thin planar membranes (which include LDPE), there is a large difference between the real volume of the swollen membrane and the volume calculated according to additive rules. Therefore, the volume and weight measurements of the swollen membranes need to be addressed in the future.

## REFERENCES

[1] Bombelli P., Howel C.J., Bertocchini F. (2017). "Polyethylene biodegradation by caterpillars of the wax moth Galleria mellonella". *Current Biology* 27: R292-R293.

[2] Khonakdar H.A., Morshedian J., Wagenknecht U., Jafari S.H. (2003). "An investigation of chemical crosslinking effect on properties of high-density polyethylene". *Polymer* 44: 4301-4309.

[3] von Pechmann H. (1898). "Ueber Diazomethan und Nitrosoacylamine". *Berichte der Deutschen Chemischen Gesellschaft zu Berlin.* 31: 2643-2646. ["About Diazomethane and Nitrosoacylamine". *Reports from the German Chemical Society in Berlin.*]

[4] Peacock A. (2000). *Handbook of Polyethylene: Structure, Properties, and Applications.* CRC Press. ISBN 9780824795467.

[5] Bowden M.E. (1994). "American Chemical Enterprise: A Perspective on 100 Years of Innovation to Commemorate the Centennial of the Society of Chemical Industry". *Chemical Heritage Foundation.* ISBN-10: 0941901130, ISBN-13: 978-0941901130.

[6] Lewis G. (2001). "Properties of crosslinked ultra-high-molecular-weight polyethylene". *Biomaterials* 22: 371-401.

[7] Kumar S., Panda A.K., Singh R.K. (2011). "A review on tertiary recycling of high-density polyethylene to fuel". *Resources, Conservation and Recycling* 55: 893-910.

[8] Hamouda H.B.H., Laiarinandrasana L., Piques R. (2007). "Viscoplastic behaviour of a medium density polyethylene (MDPE): Constitutive equations based on double nonlinear deformation model". *International Journal of Plasticity* 23: 1307-1327.

[9] Mirabella F.M., Jr., Ford E.A. (1987). "Characterization of linear low-density polyethylene: cross-fractionation according to copolymer composition and molecular weight". *Journal of Polymer Science: Part B: Polymer Physics* 25: 777-790.

[10] Hosoda S. (1988). "Structural Distribution of Linear Low-density Polyethylene". *Polymer Journal* 20: 383-397.

[11] Hadad D., Geresh S., Sivan A. (2005). "Biodegradation of polyethylene by the thermophilic bacterium *Brevibacillus borstelensis*". *J. Appl. Microbiol.* 98: 1093-1100.

[12] Yang J., Yang Y., Wu W.-M., Zhao J., Jiang L. (2014). "Evidence of Polyethylene Biodegradation by Bacterial Strains from the Guts of Plastic-Eating Waxworms". *Environ. Sci. Technol.* 48(23): 13776-13784.

[13] Chandra R., Rustgi R. (1997). "Biodegradation of maleated linear low-density polyethylene and starch blends". *Polymer Degradation and Stability* 56: 185-202.

[14] Biron M. (2018). "Thermoplastics and Thermoplastic Composites". *William Andrew*. ISBN 0081025025, 9780081025024.

[15] Friess K., Jansen J.C., Vopička O., Randová A., Hynek V., Šípek M., Bartovská L., Izák P., Dingemans M., Dewulf J., Van Langenhove H., Drioli E. (2009). "Comparative study of sorption and permeation techniques for the determination of heptane and toluene transport in polyethylene membranes". *J. Membr. Sci.* 338: 161–174.

[16] Whiteley K.S. (2012). "Polyethylene". *Ullmann's Encyclopedia of Industrial Chemistry* 29: 1-38.

[17] Barton A.F. (1975). "Solubility Parameters". *Chem. Rev.* 75: 731-756.

[18] Palatý Z., ed. (2012). "Membránové procesy". *VŠCHT Praha*. ISBN 978-80-7080-808-5. ["Membrane processes". *ICT Prague*]

[19] Izák P., Bartovská L., Friess K., Šípek M., Uchytil P. (2003). "Comparison of various models for transport of binary mixtures through dense polymer membrane". *Polymer* 44: 2679-2687.

[20] Izák P., Bartovská L., Friess K., Šípek M., Uchytil P. (2003). "Description of Binary Liquid Mixtures Transport through Non-Porous Membrane by Modified Maxwell–Stefan Equations". *J. Membr. Sci.* 214: 293–309.

[21] Friess K., Šípek M., Hynek V., Sysel P., Bohatá K., Izák P. (2004). "Comparison of permeability coefficients of organic vapors through non-porous polymer membranes by two different experimental techniques". *J. Membr. Sci.* 240: 179-185.

[22] Gholizadeh M., Razavi J., Mousavi S.A. (2007). "Gas permeability measurement in polyethylene and its copolymer films" *Materials and Design* 28: 2528–2532.
[23] Randová A., Bartovská L., Hovorka Š., Friess K., Izák P. (2009). "The membranes (Nafion and LDPE) in binary liquid mixtures benzene+methanol-sorption and swelling". *Eur. Polym. J.* 45: 2895-2901.
[24] Randová A., Bartovská L., Kačírková M., Hernández Ledesma O.I., Červenková-Šťastná L., Izák P., Žitková A., Friess K. (2016). "Separation of azeotropic mixture acetone + hexane by using polydimethylsiloxane membrane". *Sep. Pur. Technol.* 170: 256–263.
[25] Shao P., Huang R.Y.M. (2007). "Polymeric membrane pervaporation". *Journal of Membrane Science* 287: 162–179.
[26] Smitha B., Suhanya D., Sridhar S., Ramakrishna M. (2004). "Separation of organic–organic mixtures by pervaporation – a review". *Journal of Membrane Science* 241: 1–21.
[27] Randová A., Bartovská L., Hovorka Š. (2016). "Solubility of binary mixtures of n-heptane with C4 to C6 alcohols in polymeric membranes". *Sep. Purif. Technol.* 166: 213–221.
[28] Randová A., Bartovská L., Hovorka Š., Izák P., Poloncarzová M., Bartovský T. (2010). "Low-density polyethylene in mixtures of hexane and benzene derivates". *Chem. Pap. - Chem. Zvesti* 64: 652-656.
[29] Randová A., Bartovská L., Hovorka Š., Izák P., Friess K., Janků J. (2011). "Sorption of Binary Mixtures of Toluene + Lower Aliphatic Alcohols C1-C6 in Low-Density Polyethylene". *J. Appl. Polym. Sci.* 119: 1781-1787.
[30] Randová A., Bartovská L., Izák P., Friess K. (2015). "A new prediction method for organic liquids sorption into polymers". *J. Membr. Sci.* 475: 545–551.
[31] Randová A, Bartovská L., Hovorka Š., Bartovský T., Izák P., Kárászová M., Vopička O., Lindnerová V. (2017). "New approach for description of sorption and swelling phenomena in liquid + polymer membrane systems". *Sep. Pur. Technol.* 179: 475–485.

# INDEX

## A

acetone, 128, 129, 137
acid, 47, 53, 61, 68, 93, 97, 102, 105
activation energies, 23, 42
additives, 59, 60, 72, 73, 74, 76, 94, 104, 110, 111
adverse conditions, 100
agriculture, 94, 105, 122
alcohols, 122, 137
amplitude, 8, 10, 11, 13, 28, 37
antioxidant, 46, 47
antiviral drugs, 76
aquatic habitats, 86
aromatic hydrocarbons, 87, 122
Aspergillus terreus, 66, 67, 77
atmospheric pressure, 129, 131
automobiles, 85
azeotropic point, 127

## B

Bacillus subtilis, 62, 63, 78, 100, 113
bacteria, 56, 57, 58, 61, 62, 63, 65, 69, 72, 77, 78, 94, 96, 97, 98, 100, 101, 102, 110, 112, 122
bacterial cells, 112
bacterial strains, 65, 73, 97, 110
bacterium, 75, 79, 109, 112, 136
benzene, 127, 128, 129, 130, 132, 133, 134, 137
bioavailability, 102
biochemistry, 110
bioconversion, 96
biodegradability, 61, 72, 74, 77, 102, 111
biodegradable materials, 60
biodegradation, vii, viii, 56, 57, 58, 59, 60, 62, 63, 65, 67, 68, 69, 71, 72, 73, 74, 75, 76, 77, 78, 79, 86, 89, 91, 94, 97, 98, 99, 102, 103, 104, 106, 107, 108, 109, 110, 111, 112
biodegradation process, v, 55, 56, 57, 59, 61, 69, 102, 104
biological activity, 95
biomaterials, 96
biomedical applications, 32, 51
biosurfactant, 71, 78

biosynthesis, 96
biotechnological applications, 110
biotechnology, 2, 73, 76, 77, 79
biotic, 57, 59, 60, 90
biotic factor, 90
branched polymers, 24
branches, viii, 4, 16, 21, 23, 56
branching, ix, 4, 12, 17, 21, 22, 25, 32, 35, 40, 42, 43, 115, 120, 122
breakdown, 10, 90, 91, 93, 95

## C

capillary, 7, 9, 19, 30
capillary rheometers, 7
carbohydrates, 96, 97
carbon, 31, 47, 59, 64, 68, 82, 86, 88, 91, 95, 97, 103, 104, 105
carbon monoxide, 86
carbon nanotubes, 31, 47
carbonyl groups, 60, 61
carboxylic acid, 60, 97
Carreau model, 7
Carreau-like model, 23
cellulose, 75, 82, 95, 105, 107
$CH_2$, 88, 123
chain branching, 4, 21, 35, 40, 42, 43, 121, 122
chain molecules, 120
challenges, viii, 52, 56, 72, 81
chemical, viii, 4, 30, 56, 57, 62, 63, 82, 84, 85, 87, 88, 90, 95, 96, 97, 102, 107, 117, 119, 120, 135
chemical degradation, 62
chemical properties, viii, 56, 57, 63, 82, 85, 120
chemical reactions, 90
clay nanofillers, 26
comb structure, 21
commercial, 21, 24, 25, 73, 84, 85, 94, 105, 111, 119

complex fluids, 2, 11, 35
complex modulus, 9
complex viscosity, 9, 11, 19, 27, 28, 30
composites, 12, 26, 29, 30, 31, 36, 44, 75, 107
composition, 84, 95, 126, 127, 130, 131, 133, 135
compounds, ix, 27, 29, 30, 35, 64, 82, 94, 96, 104, 119, 123
consumption, viii, 81, 94, 97, 121
containers, ix, 115, 117, 121, 122
cross model, 7
crystalline, ix, 70, 115, 116, 119, 120, 123, 124, 129, 130, 131, 132
crystallinity, viii, 56, 69, 119, 120, 123, 130

## D

decomposition, 65, 68, 69, 118
deformability, 26
deformation, 2, 7, 9, 10, 11, 12, 13, 14, 15, 16, 24, 25, 135
degradation, vii, viii, ix, 56, 57, 58, 59, 60, 61, 62, 64, 66, 67, 70, 71, 72, 73, 74, 75, 76, 77, 81, 86, 89, 90, 91, 93, 94, 95, 96, 97, 99, 102, 103, 104, 106, 107, 108, 109, 110, 111, 112, 135
degradation process, viii, ix, 56, 59, 60, 62, 64, 82
degradation rate, 104, 107
depolymerization, 59, 60, 90, 91
die-swell, 19, 20, 22, 23, 40
dispersion, 28, 29, 44, 48
distillation, 127, 134
distribution, 4, 16, 17, 19, 25, 29, 41, 43, 71, 110, 121
dynamic oscillatory measurements, 8
dynamic oscillatory shear tests, 8, 10

## E

ecosystem, 57, 72, 75, 112
elongation, 16, 30, 43, 69
elongational flow, 14, 16, 24, 25, 38, 42
elongational viscosity, 14, 15, 16, 24, 25, 38
encoding, 101, 106
energy, 9, 13, 60, 64, 82, 96, 99, 127, 134
energy consumption, 127
entanglements, 18, 20, 22
environment, ix, 57, 71, 81, 82, 83, 84, 86, 88, 89, 94, 96, 99, 100, 105, 106, 108, 109, 110, 111, 122, 134
environmental conditions, 56, 89, 100, 104
environmental factors, 105
environmental impact, 51
environmental influences, 82
environmental issues, viii, 56, 61
environmental management, 107
enzymes, ix, 59, 60, 64, 75, 82, 91, 94, 96, 99, 101, 106
epidermidis, 58, 92, 98, 100
equilibrium, 8, 13, 117, 126, 127, 128, 129, 133
equipment, ix, 3, 7, 16, 115, 122
ethylene, vii, ix, 1, 4, 16, 32, 41, 51, 79, 85, 86, 115, 118, 122
exopolysaccharides, 101
experimental condition, 32
exposure, 61, 70, 79, 103, 104
external environment, 101
extrusion, 8, 14, 16, 19, 20, 23, 30, 40

## F

fillers, iv, vii, 1, 4, 26, 31
film thickness, 24, 124
films, 24, 53, 71, 72, 73, 74, 76, 88, 105, 107, 111, 121, 137
flame, 48, 49, 50, 51, 53
flexibility, viii, 56, 88
flow behavior, 2, 4, 7, 8, 11, 16, 17, 19, 21, 24, 33, 44
food, viii, 2, 52, 56, 82, 84, 88, 105, 107, 111
formation, 4, 12, 14, 27, 32, 60, 73, 78, 96, 99, 100, 101, 102, 103, 104, 106, 110, 111, 112, 113
forms, 61, 84, 86, 93, 96, 102
fungi, 56, 57, 58, 59, 61, 64, 67, 69, 73, 74, 76, 77, 94, 96, 103, 111
fungus, 73, 76, 99, 106

## G

gel permeation chromatography, 21
gene regulation, 100
graph, 67, 126, 130
graphene, 31, 45, 52, 74
graphene nanoplatelets, 31
growth, 16, 17, 24, 50, 57, 65, 67, 72, 78, 96, 97, 100, 102, 103, 106
growth rate, 57

## H

hardening, 24, 30
hardness, 87, 119
harmful effects, 91, 94
hazardous materials, 75, 77
heptane, 124, 130, 132, 136, 137
hexane, 128, 129, 134, 137
$HNO_3$, 102
homopolymers, 29, 32
household commodities, 121
hydrocarbons, 118, 119, 122
hydrogels, 33
hydrogen, 78, 82
hydrogen peroxide, 78
hydrolysis, 59, 104
hydrophilic materials, 102
hydrophilicity, 102, 103

## Index

hydrophobic properties, 102
hydrophobicity, 57, 78, 82, 102, 106

### I

ideal elastic material, 2, 8
in vitro, 65, 78, 104, 108
in vivo, 65, 108, 120
inorganic particles, 29
isothermal strain sweep test, 10

### K

kaolinite, 28

### L

large amplitude oscillatory shear, 11, 37
layered double hydroxide, 29, 44, 45, 46
LDPE-Fly ash composites, 29
LDPE-MgO nanocomposites, 30
LDPE-ZnO nanocomposites, 30
linear polymer, 3, 22, 24, 41
linear region, 10
linear viscoelastic regime, 10
long chain branching (LCB), 4, 21, 22, 23, 25, 35, 37, 38, 41, 42
loss modulus, 9, 69
low density poly(ethylene), 16, 32
low density polyethylene (LDPE), v, vii, viii, ix, 1, 4, 16, 17, 18, 19, 20, 21, 22, 23, 24, 25, 26, 27, 28, 29, 30, 31, 32, 35, 38, 39, 40, 41, 42, 43, 44, 45, 48, 55, 56, 57, 58, 59, 60, 61, 62, 63, 64, 66, 67, 68, 69, 70, 72, 73, 74, 75, 76, 77, 78, 79, 81, 82, 85, 87, 88, 90, 91, 98, 99, 104, 106, 107, 111, 115, 118, 121, 122, 123, 124, 127, 128, 129, 130, 132, 133, 134, 137

### M

macromolecular chains, 21, 26, 30
macromolecules, 4, 6, 11, 12, 14, 25, 26, 65, 96, 101, 119
materials, ix, 2, 6, 13, 14, 18, 23, 25, 31, 49, 52, 65, 74, 82, 86, 93, 100, 105, 106, 107, 111, 115, 122
melt, 2, 4, 8, 11, 14, 16, 20, 22, 26, 28, 29, 30, 31, 35, 36, 39, 40, 41, 43, 44, 46, 50, 51, 86
melt elasticity, 29
melt elongation viscosity, 30
melt flow index, 86
melt viscosity, 22, 29, 36
melting, 2, 118, 119, 123
melts, 8, 13, 30, 38, 39, 40, 43
membrane, iv, v, vii, ix, 59, 101, 115, 116, 123, 124, 125, 126, 127, 128, 129, 130, 132, 133, 134, 136, 137
membrane separation processes, 134
metabolic pathways, ix, 82, 96
metabolism, 91, 97, 101, 109
methanol, 128, 129, 130, 132, 133, 134, 137
microbial communities, 62, 94, 95
microorganism, vii, ix, 81, 92, 106
microorganisms, iv, v, vii, viii, 55, 56, 57, 58, 59, 60, 61, 64, 69, 72, 82, 91, 92, 93, 94, 95, 96, 99, 100, 101, 102, 103, 104, 105, 106, 107, 110
mineralization, 59, 60, 63, 69, 91, 95, 96, 104, 106
molar mass distribution, 17, 19, 25, 43
molecular structure, 4, 20, 25, 38, 39, 40, 41, 43, 119
molecular weight, 4, 16, 17, 18, 19, 20, 21, 22, 25, 32, 34, 36, 39, 40, 41, 42, 48, 57, 59, 60, 61, 86, 87, 94, 96, 102, 104, 106, 119, 120, 123, 135
molecular weight distribution, 4, 20, 21, 22, 25, 32, 40, 41, 120

# Index

molecules, 59, 60, 61, 64, 95, 97, 120
monomers, ix, 59, 60, 64, 82, 84, 86, 91, 104
montmorillonite, 27, 28, 44
motion dynamics, 4
municipal solid waste, 88

## N

nanoclays, 26, 27, 44
nanocomposites, 4, 14, 27, 28, 30, 31, 34, 35, 37, 38, 39, 43, 44, 45, 46, 47, 48, 52, 79
nanomedicine, 76
nanoparticles, 26, 30, 48, 49, 74, 78, 79
natural gas, 82
natural resources, 134
Newtonian behavior, 6, 31
Newtonian fluid, 2
Newtonian plateau, 6, 7, 11, 19, 23, 28, 36
non-isothermal elongational flow, 16, 25
nonlinear region, 10, 11
non-Newtonian, 2, 6, 21, 23, 33, 36, 38, 40, 41
non-Newtonian behavior, 6, 23
non-Newtonian response, 2
non-steroidal anti-inflammatory drugs, 108

## O

oxidation, 57, 60, 65, 69, 90, 94, 96, 97, 99, 102, 104
oxo-degradable LDPEs, 61, 82

## P

Pestalotiopsis microspora, 93
phenolic compounds, 47
phosphorous, 50
phosphorylation, 97
photodegradation, 90
physical properties, ix, 81, 95
physical treatments, 103
physicochemical properties, 69, 96
plastic pollution, 86, 109
plastic products, 90, 117
plastic waste disposal, 86
plastics, ix, 56, 60, 71, 76, 81, 82, 84, 85, 86, 88, 89, 90, 91, 94, 95, 97, 102, 104, 105, 106, 108, 109, 111, 112, 120, 122
pollution, 84, 86, 91, 93, 96
poly(3-hydroxybutyrate co-3-hydroxyvalerate, 93
polydimethylsiloxane, 124, 137
polydispersity, 17, 18, 19, 20, 25, 39
polyesters, 47, 74, 77
polyethylene, iv, v, vii, viii, ix, 22, 24, 28, 37, 38, 39, 40, 41, 42, 43, 44, 45, 46, 48, 54, 56, 57, 58, 59, 61, 70, 71, 72, 73, 74, 75, 76, 77, 78, 79, 81, 85, 86, 87, 88, 89, 92, 96, 102, 103, 104, 106, 107, 108, 109, 110, 111, 112, 113, 115, 117, 118, 119, 120, 121, 122, 124, 135, 136, 137
polyethylenes, 23, 24, 39, 40, 42, 43, 72, 76, 77
polymer blends, 36
polymer chain, ix, 4, 7, 11, 12, 14, 57, 82, 84, 88, 130
polymer matrix, 26, 27, 28, 36
polymer melts, 3, 12, 29, 36, 38, 39, 40, 41
polymer nanocomposites, 39, 52
polymer networks, 53
polymer properties, 94
polymer solutions, 33
polymer structure, 3
polymer systems, 2
polymer-based composites, 4
polymeric materials, 2, 13, 79, 112
polymeric membranes, 137
polymerization, 53, 118
polymerization process, 118
polyolefins, 33, 35, 71

polypropylene, 61, 71, 85, 107, 109, 119
polysaccharides, 65, 68, 101, 113
polystyrene, 36, 43, 44, 85, 86, 124
polystyrene melt, 36
polythene, 77, 78, 79, 117
polyurethane, 85, 95
polyvinyl chloride, 61, 85, 94
polyvinylalcohol, 124
polyvinylchloride, 124
power-law model, 7
preparation, iv, 35, 43
proteins, 57, 65, 68, 96, 100, 101, 105
Pseudomonas aeruginosa, 100, 106, 108, 112
pseudosolid-like rheological behavior, 28

# R

radical polymerization, 21, 122
random branching structure, 21
recycling, ix, 81, 84, 93, 122, 135
regenerative medicine, 107
relaxation, 4, 7, 11, 12, 13, 15, 18, 21, 22, 23, 25, 36, 37
relaxation modulus, 13, 37
relaxation spectra, 23
relaxation spectrum, 4, 18
requirement, 101
researchers, 57, 59, 60, 62, 65, 68, 69, 106, 124
resistance, ix, 5, 9, 26, 82, 87, 88, 100, 115, 119, 121, 122
response, 2, 4, 8, 9, 10, 11, 53, 96, 108
rheological characterization, 2, 4, 16, 25, 33, 42
rheological properties, 4, 16, 22, 29, 30, 35, 38, 39, 40, 42, 43, 44, 45, 47
rheological tests, viii, 1, 29, 32
rheology, iv, vii, 1, 2, 21, 27, 28, 29, 32, 33, 34, 35, 36, 37, 38, 39, 40, 41, 42, 43, 44, 46

rheometers, 8, 9
rotational rheometer, 7, 8, 28

# S

SAOS measurements, 18
saturated hydrocarbons, 87, 119
scanning electron microscopy, 103
separation, 26, 29, 62, 116, 123, 124, 126, 127, 128, 134, 137
shape, 19, 25, 29, 73
sharkskin phenomenon, 20
shear deformation, 5
shear flow field, 5
shear modulus, 12, 13
shear rate, 3, 5, 6, 8, 17, 20, 22, 23, 32, 36
shear rates, 3, 20, 22, 36
shear thinning behavior, 17, 23, 28, 30, 32
shear viscosity, 5, 6, 7, 8, 14, 17, 18, 19, 22, 24, 29, 36, 41
shear-thinning, 6, 7
shear-thinning region, 7
short chain branching (SCB), 22, 23
small amplitude oscillatory shear, 11
solution, 116, 117, 120, 125, 126, 127, 133
sorption, iv, vii, ix, 115, 116, 117, 123, 125, 126, 128, 130, 131, 136, 137
species, 62, 66, 71, 73, 76, 79, 98, 99, 100, 101, 107, 111, 112, 122
star structure, 21
starch, 74, 75, 103, 105, 107, 109, 111, 136
storage modulus, 9, 18, 21, 22
strain hardening, 11, 16, 24, 25, 43
stress, 2, 5, 7, 8, 9, 10, 12, 13, 14, 16, 20, 24, 29, 30, 36, 43
stress relaxation modulus, 13
stress relaxation test, 12
structure, viii, 4, 16, 21, 26, 32, 56, 65, 69, 100, 102, 120, 123
surface component, 108
surface properties, 94

swelling, 19, 116, 118, 123, 125, 130, 131, 137
synthetic polymeric materials, 71
synthetic polymers, 82, 83, 85, 89, 95, 104, 111

## T

techniques, viii, 32, 81, 91, 103, 136
technological developments, 82
technology, viii, 81, 120, 127
temperature, 8, 11, 16, 17, 21, 22, 23, 29, 30, 31, 32, 57, 62, 69, 84, 90, 93, 124, 127
therapeutic agents, 100
thermal degradation, 90
thermal evaporation, 108
thermal properties, 40, 121
thermal treatment, 102
thermo-oxidative behaviour, 47
thermoplastics, viii, 1, 2, 4, 6, 7, 9, 19, 21, 23, 26, 84, 94
thermorheological behavior, 23, 43
thinning, 6, 7, 11, 17, 18, 19, 23, 28, 29, 30, 32
tricarboxylic acid, 96, 99
Trouton law, 14, 24

## U

UV light, 60, 61, 69

UV radiation, 90, 100

## V

velocity, 5, 8, 16, 20, 30
viscoelastic behavior, 2, 8, 9, 18, 20, 22, 37, 40
viscoelastic liquids, 13
viscoelastic materials, 13
viscoelastic properties, 21, 40
viscoelastic response, 4, 11
viscosity, 3, 5, 6, 7, 8, 9, 11, 14, 15, 16, 17, 18, 19, 20, 21, 22, 23, 24, 25, 27, 28, 29, 30, 32, 36, 38, 39, 40, 41, 42, 43, 46

## W

water, 2, 28, 63, 68, 73, 74, 75, 82, 87, 91, 101, 102, 111, 119, 121, 124, 127
water absorption, 124
weight reduction, 62
welding, 120

## Z

Ziegler–Natta catalysts, 121
ZnO nanostructures, 53